Marco Zapletal
Philipp Liegl
Rainer Schuster

UN/CEFACT's Modeling Methodology (UMM) 1.0

Marco Zapletal
Philipp Liegl
Rainer Schuster

UN/CEFACT's Modeling Methodology (UMM) 1.0

A Guide to UMM and the UMM Add-In

VDM Verlag Dr. Müller

Imprint

Bibliographic information by the German National Library: The German National Library lists this publication at the German National Bibliography; detailed bibliographic information is available on the Internet at http://dnb.d-nb.de.

Cover image: www.purestockx.com

Publisher:
VDM Verlag Dr. Müller Aktiengesellschaft & Co. KG, Dudweiler Landstr. 125 a, 66123 Saarbrücken, Germany,
Phone +49 681 9100-698, Fax +49 681 9100-988,
Email: info@vdm-verlag.de

Produced in USA and UK by:
Lightning Source Inc., La Vergne, Tennessee, USA
Lightning Source UK Ltd., Milton Keynes, UK
BookSurge LLC, 5341 Dorchester Road, Suite 16, North Charleston, SC 29418, USA

ISBN: 978-3-8364-6770-4

Table of Contents

List of Figures

List of Tables

List of Listings

1 Reader

Motivation - *Philipp Liegl*

In this chapter the need for a business process modeling methodology is motivated. Furthermore a short look back into the early days of EDI is given. The transition from EDI to UMM is shown and the B2B scenario will be explained. Moreover the different chapters of the thesis are motivated.

Related Work - *Rainer Schuster*

UMM is based on some other concepts described in this chapter. Related work gives an overview about the Unified Modeling Language (UML), choreography languages, the Object Constraint Language (OCL), registries and the XML Metadata Interchange (XMI) specification.

UMM Add-In - *Rainer Schuster*

The UMM Add-In is a plug-in for the UML modeling tool Enterprise Architect. This chapter describes the communication interface between the UMM Add-In and the modeling tool. Moreover specific characteristics of the software development environment for the implementation of the UMM Add-In are described.

UMM at a glance - *Rainer Schuster*

UMM is a well accepted methodology for the modeling of interorganizational business processes. This chapter shortly describes the artifacts and purposes of the three main views of UMM. Furthermore the tasks of the permanent working groups of UN/CEFACT and the organizational structure are explained.

Worksheets Editor - *Rainer Schuster*

UMM defines a set of worksheets that are used for the communication between business domain experts and business process analysts. These requirement documents are usually created by word processors and are separated from the UMM model. Most of the information captured in work-sheets is represented one-to-one in tagged values of the model's stereotypes. Thus the same information is hold twice. The management of the redundant information does not only result in expensive efforts, but also results in a high danger of inconsistency. In order to overcome these limitations, the worksheets must be integrated to the modeling tool itself. In this chapter we present the concepts and the implementation of an interactive worksheet editor. This worksheet editor allows the dynamic binding of worksheet elements to UMM modeling elements by using a special worksheet definition language. This definition language also guarantees a flexible adaptation of worksheets to special business needs and to changes due to the update of the UMM meta model.

User Guide - *Marco Zapletal*

This chapter presents a guide for creating UMM compliant process models. It depicts the complete workflow of creating a business collaboration model by detailing the modeling tasks of each view and each subview of the UMM. Furthermore, we discuss how the input from business experts gathered by worksheets results in corresponding UMM model structures. Each step in this chapter is illustrated via an example describing an ordering scenario.

UMM Validator - *Philipp Liegl*

The chapter about validation will give an overview about the different extension mechanisms which are provided by UML and how they are used within the UMM. The conceptual UMM model is explained and an overview about the OCL constraints which are used as the basis for the validator is given. Furthermore the UMM validator and its architecture will be explained.

Deriving Process Specifications from UMM Models - *Marco Zapletal*

In modern service oriented architectures, choreography languages are utilized to specify processes in a machine-executable manner. In order to avoid writing such process descriptions by hand a model driven approach is desired. Since UMM business collaboration models capture collaborative processes, the desire for an automatic generation of corresponding choreography descriptions from UMM models is self-evident. In this chapter we describe a mapping from UMM collaborations to process descriptions written in the Business Process Execution Language (BPEL). The mapping is described on the conceptual and on the implementation level.

Mapping Business Information to Document Formats - *Philipp Liegl*

This chapter stresses the importance of business information modeling. The concept of Core Components (CCTS) will be introduced. In the next step the Universal Business Language (UBL) will be explained, which builds on the CCTS. Furthermore the need for naming and design rules for XML documents is be discussed. At the end of the chapter a reference implementation within the UMM Add-In is shown.

Summary and Outlook - *Marco Zapletal*

This chapter gives a short conclusion in respect to this thesis as well as in regard to the implementation of the UMM Add-In.

Appendix - *Philipp Liegl, Rainer Schuster, Marco Zapletal*

This chapter comprises the business transaction patterns and the bibliography.

UMM Foundation Module 1.0, Technical Specification

The UMM Foundation Module 1.0, Technical Specification, which we co-authored during the work on our thesis is available online at http://www.unece.org/cefact/umm/umm_index.htm

UMM Add-In 1.0

The UMM Add-In, which we implemented during the work on our thesis is available for free at http://ummaddin.researchstudio.at/

2   Motivation

Fourty years ago the average enterprise was not equipped with IT infra-   *The evolution of business*
structure as we know it today. Business processes were processed between   *processes*
enterprises by using common postal correspondence or by bilateral agreements between trading partners - basically there was no need to synchronize
or orchestrate any IT processes between the enterprises. A typical scenario
would probably have been the following:

> Enterpise A produces good x. Enterprise B buys good x from Enterprise A and uses
> it to produce its own good.

The only choreography which took place was the purchasing department of
enterprise B calling the production department of enterprise A, making sure,
that punctual and appropriate delivery of good x takes place. Nevertheless
already note, that both enterprises are stakeholders in the production/deliv-
ery process. Enterprise A wants to produce good x in order to sell them to
Enterprise B and Enterprise B needs good x in order to produce its own
good.

Today such a collaboration takes place between numerous enterprises
in all fields of the economy. As more and more enterprises are using the IT
as a production factor and crucial supporting factor to their internal processes a choreography of the collaboration is necessary. Almost every collaboration involves software which needs to be developed, customized or
orchestrated. The UN/CEFACT's Modeling Methodology (UMM) helps to
capture the business knowledge, which is necessary in order to develop low
cost software which can help small and medium size companies to engage
in e-business practices.

This thesis will focus on the UMM standard and its tool based support
by the UMM Add-In. We have developed the UMM Add-In conjointly with
this thesis. The Add-In is a software extension for the UML modeling tool
Enterprise Architect and helps the modeler in achieving a valid UMM
model. Before we start to immerse into the UMM Add-In and its development and functionality we will have a look at the development of business
processes between enterprises and the standards and efforts which have
been made. We start with an overview about B2B development in the past.

The reader of this thesis must have a deep understanding of UML 1.4 [UMa04] and must be able to understand meta models denoted as UML class diagrams. He should be familiar with the UML 1.4 meta model, at least he must be able to check back with the UML 1.4 meta model. As UMM [FOU03] is the basis for this thesis, a basic knowledge of the standard is helpful. Furthermore a basic understanding of Enterprise Architect from Sparx Systems is advantageous.

Requirements for the audience

2.1   B2B - a look into the past

Going back 30 to 40 years from now only few enterprises had an IT department and even fewer had a network connection. B2B processes were rather unknown or at least IT professionals understood a different thing in regard to business to business processes. The concept in the field of B2B which has been developed since the mid-1960s was the Electronic Data Interchange (EDI). We will now have a look at the historical development of this data centered approach.

2.1.1   Electronic Data Interchange

The main aim of Electronic Data Interchange (EDI) is to eliminate paper documents for the exchange of business data. The first attempts in this direction were already made at the time of the Berlin Airlift [Sch88].

The first data centered standardization approaches

Yet the history of EDI began when transportation data which was exchanged between companies of a railroad group in the United States was lacking the expected quality. Therefore a group was found which should study the problem and increase the data quality and the feasibility of its electronic exchange. It was know as the Transportation Data Coordinating Committee (TDDC). At the same time companies in the automotive sector addressed a similar issue by developing their own proprietary data exchange systems. These systems were able to exchange electronic data with the major trading partners of the automotive industry. However due to a missing universal standard a potential trading partner was supposed to have a different system interface for every trading partner with a different EDI system.

One of the first companies to detect the need for a industry specific standard was from the grocery industry. The company had to handle large interorganizational EDI issues and was therefore eager to develop a standard specific to an industry, namely the grocery industry. Nevertheless an approach for a universal standard was not pursued because it was considered to be unnecessary and not practical for the technology levels which were available back then.

In the 1970's the next step by several industries was sponsoring a shared EDI system. The idea was turning over the data exchange system to a third party network. In some cases a third party even developed a shared system for a group of common companies. However these industry trade group systems were encountering the same limitations as the first EDI systems. They were limited in scope and unable to communicate between different sectors of the industry and their respective EDI systems.

Hence in 1973 a set of standards for EDI between companies was invented. The initiative was coming from the Transportation Data Coordinating Committee which wanted to have a standard that was able to react on changes in the requirements. Therefore the standard included procedures on how to change the standard as well. This first inter-industry EDI standard covered the air, motor, ocean, rail and some banking applications.

The most important step was taken, when in 1985 the work started on UN/EDIFACT (United Nations/Electronic Data Interchange for Administration, Commerce & Transport) supervised by the United Nations.

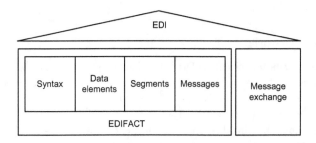

Fig. 2–1 The components of the UN/EDIFACT standard

Figure 2–1 gives an overview about the UN/EDIFACT standard. The syntax defines the rules for the definition of a message structure. The standard furthermore includes standardized codes and data elements which can be reused. Groups of data elements which belong together can be aggregated in so called segments. Messages represent a structured sequence of segments.

Listing 2–1 shows a part of an arbitrary UN/EDIFACT message. As one can see, the syntax is quite complicated and almost impossible to understand for the humans.

```
[1]   PDI++C:3+Y::3+F::1'
[2]   APD+74C:0:::6++++++1A'
[3]   TVL+240493:1740::2030+JFK+MIA+DL+081+C'
```

Listing 2–1 A piece of an arbitrary UN/EDIFACT message

UN/EDIFACT itself is a text based exchange format which is platform independent. With the rise of the eXtensible Markup Language (XML) which is another platform independent format with mark-up ability, efforts have been made to combine XML and EDI. Such an approach has been pursued by the XML/edi Group. The main aim of the group was not only combining XML and EDI but widening the standard with additional templates, agents and repositories. The idea is, that electronic systems should not only exchange business data but provide templates for the processing of the data and business rules as well. However the transformation of UN/EDIFACT to XML has proven to be difficult because the transformation of the semantics of a UN/EDIFACT message into a XML representation is not unambiguous.

As Listing 2–1 already showed, UN/EDIFACT is a quite complex standard which consists of numerous data elements, segments etc. Its implementation and use requires the use of experts. Small and medium sized enterprises often cannot afford such manpower.

In 1999 an initiative was started by UN/CEFACT and OASIS for the standardization of an XML specification for electronic business. It was called ebXML (Electronic Business using eXtensible Markup Language). Its main aim was the creation of an open technical framework for the exchange of business data with regard to interoperability and affordability especially for small and medium enterprises. XML messages should support businesses by providing a standardized document structure. Furthermore the choreography of business processes should be defined. Another aim of ebXML is the use of standardized collaboration protocol agreements between business partners under the use of commercial of the shelf software (COTS).

ebXML as a new approach to EDI

The next chapter will show how the transition from EDI to UMM was accomplished and which problems occurred.

2.1.2   The transition from EDI to UMM

After decades of data centered endeavors with UMM finally a process centered standard was developed. As already shown in the last paragraph, EDI concentrates solely on the data which is exchanged between business partners and does not take into account the choreography necessary for the business process alignment. With ebXML a substantiated framework was found, which supports the choreography of the business processes and the exchange of business messages.

However UMM does not provide a standard for the exchange of messages as it is done by ebXML. UMM is process centered and focuses on the choreography of collaborative business processes. Nevertheless it is also possible to specify the information, which is exchanged between business partners. UMM does not set a specific data model for the information to be exchanged but leaves this decision open to the implementor. Hence different standards like UN/EDIFACT, UBL etc. can be used. Therefore the UMM

From a data centered to a process centered approach

modeler does not have to know the rather difficult exchange standard as for example UN/EDIFACT but can exclusively focus on the modeling of the collaborative business process. Furthermore the business context of the collaborative process is captured in a syntax neutral manner.

The business logic and the technology which implements the logic are separated with UMM. Hence UMM is a model driven approach which focuses on the business processes that take place between two collaborating parties. Furthermore it focuses on the business state of business entities and on the business context of the business process. The current UMM standard is defined in [FOU03] that provides the basis for this thesis.

We will now examine the B2B scenario which is the basis for a UMM model. Shortcomings of old approaches and enhancements through the usage of UMM will be shown.

2.2   The B2B scenario

When talking about business process modeling today, most people will refer to the modeling of business processes which are internal to an organization. This so called business process management mainly focuses on the optimization of business processes and the integration of enterprise applications within a company.

However because today business processes also take place between organizations, the focus must change from the intra- to the interorganizational view. Today most companies have accomplished this step and are therefore not only centered on their own processes anymore.

In order to allow two companies to collaborate, a choreography has to be defined which choreographs the business processes between the companies. Unfortunately if each organization defines its own choreography with each business partner it is very unlikely that interoperability can be achieved.

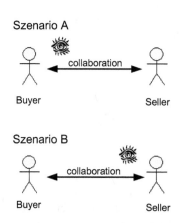

Fig. 2–2 Different views on a choreography

Figure 2–2 shows a sample collaboration between two parties. In scenario A *buyer* has defined his own choreography for the collaboration with *seller*. In scenario B *seller* has defined his choreography for the collaboration with *buyer*. A collaboration between *seller* and *buyer* is highly unlikely because each business partner has defined its own choreography. A mechanism would be necessary, which allows the development of collaborative business processes and information models between two business partners in an easy and protocol independent manner.

At this point UMM is the modeling methodology of choice. It allows to capture the business knowledge necessary for companies in order to support collaborative business processes. Figure 2–3 shows a collaborative business process using UMM.

Fig. 2–3 A collaboration scenario using UMM

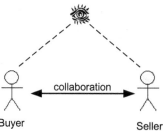

By using UMM a choreography can be found which supports a collaboration between *buyer* and *seller*. Furthermore the overall structure for the information to be exchanged during the business process can be defined as well. The information modeling and transforming issue will be addresses in more detail later.

2.3   The need to define interfaces

When talking about a collaborative business process the mentioning of the term interface is inevitable. Modern business processes typically include IT systems which have to communicate - for instance they exchange business documents which are stored in interchange files.

Large companies usually use enterprise software which has been exclusively developed for that specific company. Small and medium sized companies cannot often not afford expensive enterprise software and are therefore forced to use low cost commercial of the shelf software (COTS). Business information is usually exchanged by using commonly known document type formats which COTS systems understand as well. However an additional agreement has to be set between two business partners in order to customize the standard document type to the partnership specific requirements. Therefore the business partners have to customize their import and export functionalities for the partnership specific requirements. Such an adaptation is usually no problem for a large enterprise which runs its own personalized software.

However for small and medium sized enterprises an adaptation of their software is almost impossible. Therefore for the time being only large companies can perform such B2B processes. Small and medium sized enterprises need the business functions and the B2B functionality integrated into their COTS. To believe, that COTS software vendors will agree on a single data requirement for a particular document type is unrealistic. *The disadvantage of small and medium sized companies*

Hence another approach has been proposed. Research was conducted by UN/CEFACT in order to look for alternatives to the scenario mentioned above. The development of a well defined business process for each particular business goal was proposed e.g. *order from quote*. Such a well defined business process contains all possible activities which could be part of the business goal. Because the business processes are collaborative processes we refer to them as business collaborations. There can be many ways in which a business collaboration can be executed. However every one is well defined. It now depends on the trading partner and his internal processes, which alternative he can execute. One trading partner might be able to execute all alternatives whereas another trading partner might only be able to execute a few. However in order to start a collaborative business process, the two business partners must be able to engage in at least one alternative, which they have in common.

As that is no problem for the large enterprises, small and medium sized enterprises rely on software providers who should create applications which implement business collaborations with their most popular execution scenarios. In order to pursue such an approach unambiguous business collaboration models in regard to choreography and involved document structures are required.

Our thesis will show how UMM is able to fulfill the criteria of finding business collaboration models.

2.4   The need for tool support

Together with this thesis an Add-In for Enterprise Architect was created, that supports the UMM modeler in creating a valid UMM model. In this chapter we would like to motivate the need for such a tool.

First of all the decision to take Enterprise Architect and not a modeling tool from another vendor was not an arbitrary one. Enterprise Architect offers an interface which can be easily accessed by any language which is able to access Microsoft Windows COM components. Hence Enterprise Architect together with Microsoft Visual Studio and C# was chosen.

Enterprise Architect as the tool of choice

When the modeler starts to create a new UMM model an initial UMM structure must be created. An automated creation of such an initial model structure would accelerate the model creation and allow beginners in UMM to start with a basic settlement.

Before a UMM model is created which describes a collaborative business process, the business knowledge has to be collected first. Business knowledge is usually collected during interviews with business experts and computer engineers which is then written down in plain text. Before the UMM Add-In this information collection process was done manually by using Microsoft Word documents to write down the collected information. For instance descriptions for business processes were stored in Microsoft Word documents while the actual business process was modeled in a modeling tool. Seen from the point of usability this separation is a disadvantage in regard to distribution and reuse of a model. Storing the information concerning a specific business process or a specific process all together would be a great enhancement.

Capture the requirements

As later chapters will show, business processes can be categorized by using business categories, business areas and process areas. The alignment of these categories and areas follows a certain pattern according to the specific area of the industry. If the user would be provided with an industry overview where he could choose his specific area together with an automated generation of the specific packages within the modeling tool this would optimize the modeling workflow during the requirements phase.

Categorize the business processes

As we will show, the UMM meta model can be quite complicated for an inexperienced user. Hence one can suppose, that a lot of modelers will be anxious to use the UMM modeling standard for their business process modeling. A tool support which allows the modeler to check whether the created

Validate the created UMM model

model is valid or not would alleviate the modeling process for the inexperienced users and lower the threshold for those unsure to use UMM. However also experienced UMM modelers could use a validation functionality in order to scrutinize a created model before further use.

A further use could for instance be the generation of choreography languages from a UMM model. As already shown in chapter 2.1.2 UMM focuses on the choreography and orchestration of business processes. For the time being two significant choreography standard exist namely Business Process Specification Scheme (BPSS) and Business Process Execution Language (BPEL). A great benefit to the modeler would be the ability of automatically generating such choreography standards from a UMM model.

In regard to user interface design Enterprise Architect has some shortcomings, which hamper the modeling workflow. UMM is a standard which thoroughly uses the concept of *stereotypes*. Some of the used stereotype names are quite long as for instance *BusinessCollaborationUseCase*. If the user wants to change the stereotype of an arbitrary *use case* to *BusinessCollaborationUseCase* this is currently not possible with Enterprise Architect. The tool only allows a certain number of letters for the name of a stereotype. However if the assignment of the stereotype is done programmatically via the API of Enterprise Architect (which the UMM Add-In uses) any length is possible for the name of a stereotype. Hence the implementation of a so called "Stereotyper" which allows to assign stereotypes of any length would enhance the modeler's workflow.

All the stimuli mentioned above have been implemented by the UMM Add-In and will be explained in more detail throughout this thesis. The next chapter within the motivation for this thesis is the need for a user guide, which helps a modeler to create a valid UMM model.

2.5   The need for a user guide

Even when provided with the UMM meta model and a basic understanding of UML it has proven to be quite difficult for a modeler to create a UMM compliant model from scratch. A guide is needed, which takes the modeler from the beginning to the end of the modeling process and provides additional information to the specific subviews and subpackages of a UMM model. The UMM meta model defines the constraints for a valid UMM model but it does not show the modeler how to build a valid model. This gap can be closed by a user guide.

The meta model on its own is not enough

It would be helpful for the modeler if every package is described and if an overview about the packages and their purpose is given. Furthermore the modeler might ask himself, what the stereotypes exactly mean and which stereotypes are to be used in a given package. After having explained the details for a specific package a step by step modeling guide would help the modeler to create packages valid in the sense of the UMM meta model. Together with screenshots a step by step modeling guide would lead the modeler towards a valid UMM model.

A user guide as described above is included in this thesis.

2.6   The need for business information transformation

As already outlined before, UMM focuses on the modeling of collaborative business processes between business partners. A business process usually involves the exchange of information as well. Further chapters will show that within UMM several B2B document standards can be used in order to model the information exchanged during a business process.

Because UMM is a graphical modeling methodology, the information which is exchanged will be modeled graphically. Several components of an information will stay the same for different implementations. An *address* will for instance most likely contain a *state* and a *post code*. Using such reusable components for the information modeling would alleviate the modeling process. As we will see within this thesis, the usage of Core Components Technical Specification (CCTS) will enhance the information modeling and facilitate the reuse of information modeling components.

Reuse of information modeling components

Current research is undertaken in order to directly map the business information modeled graphically into an exchange format. As a method of choice we will present the transformation of business information into an XML schema. The XML schema derived from the information model serves as a normative reference for all XML document instances. Hence an unambiguous document format can be guaranteed.

2.7   The motivation for this thesis

As the introduction has already shown, a lot of theoretical work has been done in the last few years in the field of collaborative business process research. Different approaches have been evaluated and a lot of paper have been produced. However there exists no practical solution which supports the modeler in creating a valid UMM model. A lot of work is currently done by the working groups of UN/CEFACT in order to accomplish the goal of a common business process modeling methodology. Without a specific software implementation the efforts to create and diffuse a standard are not likely to succeed.

The UMM Add-In and this thesis support modelers in creating a UMM compliant model. It serves experienced UMM professionals as a reference, medium skilled UMM modelers as a support and UMM beginners as a first step towards a correct UMM model. We hope to support the diffusion of UMM as a basic standard for the modeling of collaborative business processes.

3   Related Work

3.1   UML

The Unified Modeling Language (UML) is a graphical language for visual-   *The purpose of UML*
izing, specifying, constructing, and documenting the artifacts of a software-
intensive system [UMa04]. UML offers the possibility for designing and
implementing software models in a normative way under consideration of
object-oriented structures. UML is a standard since 1998 and is maintained
by the Object Management Group (OMG). With this standard a modeler can
create conceptual models including business processes and system functions as well as concrete models such as programming language statements,
database schemes and reusable software components. The modeling language was developed by Grady Booch, Ivar Jacobsen and Jim Rumbaugh
[BRJ05] [BRJ04] from Rational Rose Software.

The specification of UML defines a set of different diagrams [UMb04].   *The artifacts of UML*
Each diagram provides a different perspective of the system under analysis
or development. The following list shows the primary artifacts of UML representing eight different diagrams:

- Use case diagram
- Class diagram
- Behaviour diagram
    - Statechart diagram
    - Activity diagram
    - Interaction diagram
        - Sequence diagram
        - Collaboration diagram
    - Implementation diagram
        - Component diagram
        - Deployment diagram

UML defines a set of notations for creating a specification language for   *UMM is a UML profile*
software development. UML is not a method. For using UML efficiently the
task is to develop a method which is UML compliant. Such a method e.g. is

UN/CEFACT's Modeling Methodology (UMM). UMM is a specific UML Profile for modeling collaborative business processes [UG03].

Since the UMM meta model uses the UML version 1.4.2, all concepts described in this thesis are based on the UML version 1.4.2 [UMa04] [FOU03]. The latest UML version is UML 2.0 [UMb04] and is supported by most of the current UML tools including Enterprise Architect from Sparx System.

The different versions of UML

3.2   OCL

The Object Constraint Language (OCL) [OCL03] is a formal language for describing constraints on Unified Modeling Language (UML) models. UML is using different modeling rules which have to be considered during the modeling process. Since a modeler can define new profiles basing on these UML rules, he must be able to put restrictions on this new modeling method. Such restrictions are described in OCL.

OCL describes constraints for UML

Usually UML diagrams are not specified enough by adding comments in a natural language. The semantic of such text-based descriptions could be understood variably. The OCL supports a possibility for capturing all the relevant aspects of a specific UML diagram in a normative way. Thus the modeler can add constraints to any part of the UML profile to ensure, that the user does not cause any ambiguities.

OCL prevents ambiguities

OCL fills the gap of the semantic problems with a so-called formal language, which is easy to read and write and has an unambiguous meaning. It has been developed as a business modeling language within the IBM Insurance division, and has its roots in the Syntropy method during the early 1990s [Ste94]. OCL should not be mixed up with a programming language, because an OCL expression can not cause side effects. This means after executing an OCL constraint, only a value with the information about the validation success of the UML construct is returned. Thus OCL cannot change anything in the model and can not implement programming logic or flow control. OCL defines a so-called standard library for a set of supplementary predefined OCL types. Since OCL is a typed language, each OCL expression has a type. For creating a well formed OCL expression the type of the expression must conform to the type conformance rules of the language. This restriction ensures that you can not compare e.g. an *integer* with a *string*.

Filling the gap of semantic problems

OCL can be used for different purposes:
- As a query language
- For specifying *invariants* (special OCL type) in the class model

Different purposes of OCL

For describing *pre-* and *post-conditions* on operations and methods

For describing guards

For specifying target (sets) for messages and actions

For specifying constraints on operations

For specifying derivation rules for attributes for any expression over a UML model

In this thesis the Object Constraint Language is used to define the constraints on the UMM meta model. These OCL definitions became part of the UMM Foundation 1.0 specification.

3.3 Choreography languages

UN/CEFACT's Modeling Methodology (UMM) is a methodology for describing interorganizational collaborative business processes. A UMM compliant business model concentrates on the business semantics of the business processes. This concept is called *business operational view* (BOV) and is a part of the Open-edi reference model [OER95]. This semantical part of UMM must be supported by an IT infrastructure of the *functional service view* (FSV), which is responsible for the execution of business collaborations. Thus a choreography is used to describe the resulting interorganizational process from different perspectives. Since the resulting process is carried out by software modules, the definition of choreographies must be machine-readable. For this reason the formats of such specifications are usually XML-based.

The task of a choreography language

Talking about the execution of collaborative business processes, we have to make a distinction between two terms – *choreography* and *orchestration*. A choreography is the relation between business processes in a peer-to-peer collaboration trying to reach a common goal. It tracks the business document exchanges among multiple parties and sources. Consequently the perspective of a choreography has a collaborative nature without describing internal tasks [Pel03]. The second term describes the perspective of an orchestration. An orchestration is the sequence in which one business process invokes other business processes for reaching a goal [WSG04]. This process is executed within the boundaries of an organization. Such interactions are at the message level, including the business logic. In this thesis both terms are related to the definition of an execution sequence. Since UMM is used to model collaborative business processes, it mainly deals with choreography.

Choreography vs. Orchestration

As already mentioned the business process model must be transformed to a machine-readable language. What else is more suitable for this task as

Different choreography languages

XML. Thus the following list shows the most important XML-based chore-ography languages [Hof05]:

- ebXML Business Processes Specification Schema (BPSS) [BPS03]
- Business Process Execution Language (BPEL) [BEA03]
- Business Process Modeling Language (BPML) [BML02]
- XLANG [XLA01]
- Web Services Flow Language (WSFL) [WSL01]
- Web Services Choreography Interface (WSCI) [CI02]
- Web Services Conversation Language (WSCL) [CL02]
- Petri Net Markup Language (PNML) [Kin04]
- Event-Driven Process Chains Markup Language (EPML) [MN04]
- Graph eXchange Language (GXL) [WS04]

Due to this large number of different choreography languages there is a short description of the most commonly used choreographies.

The Business Process Specification Schema (BPSS) defines a frame-work for business process specification. Its goal is to provide the bridge between e-business process modeling and the specification of e-business software components. The Business Process Modeling Language (BPML) is a block-structured meta language. It is based on a logical process model expressing concurrent, repeating and dynamic tasks and can be directly exe-cuted via middleware. The viewpoint of a single partner is a main character-istic of this choreography language.

Description of the most important choreography languages

The Business Process Execution Language (BPEL) is a combination of the concepts of XLANG and WSFL specifications. BPEL uses so-called *business protocols* for sequencing the messages exchanged by business partners. Since a business protocol is not executable, BPEL refers to it as an *abstract processes*. They are meant to couple Web Service interface defini-tions with behavioral specifications that can be used to both constrain the implementation of business roles and define in precise terms the behavior that each party in a business protocol can expect from the others [Jur04]. In other words this is a process interface from the point of view of the party exposing this interface. In order to realize a business protocol BPEL uses the concept of *executable processes*. Executable processes extend abstract processes by defining concrete protocol bindings and the exchanged docu-ment structures.

The Business Process Exe-cution Language (BPEL)

3.4 ebXML

The United Nation's Centre for Trade Facilitation and e-Business (UN/CEFACT) and the Organization for the Advancement of Structured Information Standards (OASIS) started an initiative called ebXML. The goal of ebXML is the provision of an XML-based infrastructure for a world-wide usage of electronic business processes. The ebXML project was founded in November 1999, because a global electronic business market was needed. In this global market businesses can find each other to become trading partners in a cost-efficient way. This concept bases on the exchange of XML documents. Primarily small and medium enterprises (SME) should get advantages out of this new initiative. The software industries will offer commercial off the shelf software (COTS) to the SMEs for realizing B2B scenarios.

ebXML - an XML-based infrastructure for a world-wide usage of electronic business processes

Fig. 3–1 An ebXML scenario

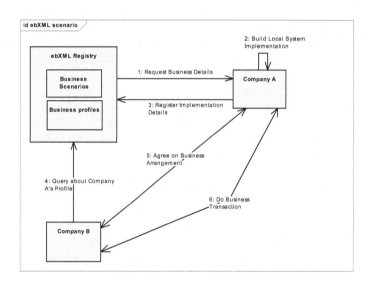

Figure 3–1 depicts a typical ebXML scenario between a large company (*Company A*) and an SME (*Company B*) [ETA01]. There is a central communication interface called ebXML registry. In the first step *Company A* is requesting information from the ebXML registry. For submitting their own business process information *Company A* has to implement their own ebXML-compliant application. The third step is registering the implementation details to the registry and submitting the profile of *Company A*. This profile gives information about the company's ebXML capabilities and constraints, and its supported business scenarios. *Company B* sends a query

Description of an ebXML scenario

about the profile of *Company A* (step 4). Having all the required information, *Company B* agrees on the business arrangement in the fifth step. Since both companies are using an ebXML compliant software interface, they can directly communicate with each other. *Company A* accepts the business agreement as well. Now the e-business process using ebXML is starting in the sixth step.

3.5   Registries

The idea of a registry is keeping information and methods accessible for specific participants connected to the Internet. This scenario is useful for business processes as well. As described in chapter 3.4 ebXML uses this concept for keeping their business processes in a central communication interface. Therefore the characteristics of a registry will be explained on the basis of an ebXML registry.

The task of the ebXML registry

The ebXML registry builds a repository for registering business profiles. An ebXML registry can be compared with a relational database of enterprise applications. The following informal definition supports this statement: an ebXML Registry is to the web what relational databases are to enterprise applications [Naj02].

An informal definition of the ebXML registry

Enterprise applications are evolving into web services with the need for sharing data and metadata. An ebXML registry may be used by web services and applications to store and share content and metadata. A more formal definition of the ebXML registry is as follows: An ebXML registry is an information system that securely manages any content type and the standardized metadata that describes it [Naj02].

A more formal definition of the ebXML registry

Moreover the ebXML registry provides services maintaining the shared information as objects in a repository. The information stored in the registry could be collaboration protocol profiles of trading partners, core libraries, business libraries, business processes or business documents [Hof05]. All these artifacts are managed by the ebXML registry.

Artifacts managed by the ebXML registry

Talking about registries we must consider the term UDDI (Universal Description, Discovery, and Integration). UDDI is a web-based distributed directory that enables businesses to list themselves on the Internet. It is an XML-based registry which allows businesses worldwide to discover each other. Its ultimate goal is to streamline online transactions by enabling companies to find each other on the Web and make their systems interoperable for e-commerce. In order to find the right process for your company UDDI can be compared with telephone books for different purposes - white pages, yellow pages, and green pages. Furthermore the information in the UDDI registry provides a mechanism that allows others to discover what technical programming interfaces are provided for interacting with a business.

The purpose of UDDI

The information that a business can register includes several kinds of simple data that help others determine the questions "who, what, where and how". The question "Who?" is answered by the information about the business such as name, business identifiers and contact information. "What?" involves classification information that includes industry codes and product classifications, as well as descriptive information about the services that business makes available. The question "Where" involves registering the information about the URL or email address (or other address) through which each type of service is accessed. Finally the question "How?" is answered by registering references to information about interfaces and other properties of a given service. These service properties describe how a particular software package or technical interface functions [UDD02].

The data in UDDI determines the questions "who, what, where, and how"

3.6   XMI

The XML Metadata Interchange (XMI) is the combination of two major buzzwords: UML and XML. Although UML is a normative language, the portability between different UML tools is a problem. There are a lot of UML tools on the market. Bringing the same information out of a UML tool requires a normative description of UML diagrams in a formal language. Since different UML tools are recording different information, this was the major reason for inventing an XML-based exchange format for UML models [Ste01]. Thus XMI is a way to save UML models in XML. The XML Metadata Interchange is an OMG standard. The latest version of XMI is 2.1 and is specified in the MOF (Meta Object Facility) 2.0/XMI Mapping Specification v2.1 [MOF05]. UML is a MOF-based meta model and therefore XMI should be specified in a more general definition [MO102]. XMI shows how to save any MOF-based meta model in XML.

The task of XML Metadata Interchange (XMI)

Before going on with the description of XMI, the Meta Object Facility (MOF) must be described in more detail. The MOF is an Object Modeling Group (OMG) standard such as XMI. MOF is a simple language for defining further languages, such as UML. The OMG specifies a 4 level meta model architecture for defining MOF. These different layers are depicted in Figure 3–2. The architecture consists of 4 different meta model layers. The *M3* meta level is the meta-meta model. The MOF model itself is an example of this layer. The *M2* layer keeps the meta model. An instance of this layer is the UML meta model defining the abstract syntax for the relationships between any kinds of UML model elements. The model itself is at the *M1* level. Every UML model can be used as an example of this layer. The *M0* level keeps the data of a modelled system. An important fact of this 4 level architecture is that MOF must not be mixed up with UML. It is just providing an "open-ended information modeling capability".

The Meta Object Facility (MOF)

*Fig. 3–2 OMG's 4 level
meta model architecture*

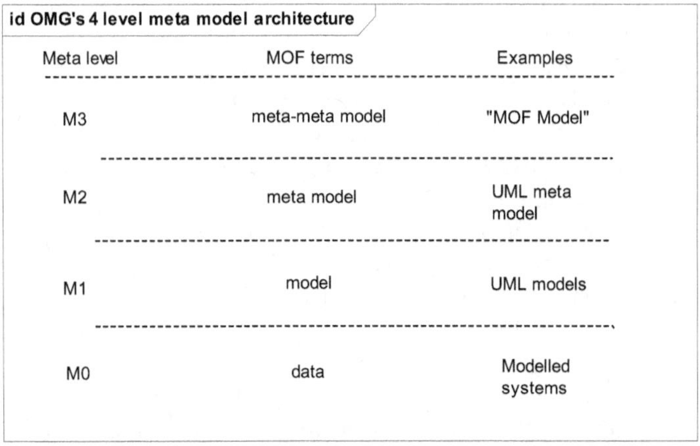

Meta level	MOF terms	Examples
id OMG's 4 level meta model architecture		
M3	meta-meta model	"MOF Model"
M2	meta model	UML meta model
M1	model	UML models
M0	data	Modelled systems

As we can see UML models can be transformed to XML-based formats using the rules of XMI. The XMI format is geared to the UML meta model (M2). The framework of the tags in the XMI refers to the structure of the UML meta model. The data within the tags reflects the UML model. Thus XMI is an exchange format for UML models. Since the UML meta model is integrated in every UML tool, the XMI format must be compliant to each UML tool.

XMI is an exchange format for UML models

Unfortunately the majority of UML tools do not implement the UML meta model itself correctly. Hence these UML tools export their UML models in different ways using different output flavors. Furthermore some UML tools do not implement specific information of UML models which may be required by other UML tools. Thus the interoperability of XMI between UML tools is a major issue.

Although these problems may show a lot of disadvantages of XMI, there are some helpful aspects of this exchange format. Since XMI is XML based a broad range of tools are available to manipulate models persisted in XMI. Furthermore a developer can write code for extracting information out of an XMI file in order to transform it into the input format of another UML tool.

The advantages for UML using XMI

If there were no problems concerning the interoperability of XMI formats between different UML tools, we would have considered using the concepts of XMI for the UMM Add-In. The UMM Add-In includes a feature for validating a UMM model against the constraints defined in the UMM meta model. In the beginning of the development phase it was planned to use XMI version 1.2 for validating the exported XMI file of a

Why does the UMM Add-In not use XMI?

UMM model. The UMM Add-In is a plug-in for Enterprise Architect, which supports a multifunctional application integration interface. Using the interface for the validation is more comfortable than using XMI. Since these XMI files cannot be used in other UML tools anyway, for interoperability reasons the application integration interface is used instead. Furthermore it saves a lot of program code, because the information of the UMM diagrams does not need to be extracted to an XML-based format. Thus the idea of using XMI was rejected.

4   UMM Add-In

4.1   An Add-In for a UML modeling tool

Since UMM is based on the concepts of the Unified Modeling Language *An overview about UML*
(UML), a tool is needed for modeling a UMM business collaboration *modeling tools*
model. There are a lot of UML tools on the software market. The most
known tools are Rational Rose from IBM, Enterprise Architect from Sparx
Systems, Magic Draw from NoMagic Inc., and Poseidon from Gentleware.
The modeling tools provide UML model elements defined in the UML meta
model. Furthermore these software tools can be usually used to load a UML
Profile - e.g. as it is defined for the UMM. A stereotype is a capability to
create new kind of modeling elements based on elements that are part of the
UML meta model. A stereotype can be customized by defining attributes
called *tagged values*. UMM stereotypes are defined in the UMM meta
model which is UML compliant as well.

 For supporting the modeler in building a UMM model, a tool which *The purpose of the UMM*
guides through the modeling process is needed. The main tasks of this *tool*
UMM tool are:

- Increasing usability of the modeling process
- Providing UMM compliant stereotypes
- Documentation of the model
- Validation of the model
- Transformation of the model into choreography languages
- Mapping business information to specific XML based document formats

The best way to fulfill these tasks is an integration of the UMM tool into a *The Enterprise Architect*
UML modeling tool. Thus we chose the UML tool Enterprise Architect *supports an interface for*
(EA) from Sparx System for creating a plug-in called UMM Add-In. Enter- *integrating the UMM Add-*
prise Architect provides an API for using predefined features and methods *In*
of the modeling tool.

4.2 Enterprise Architect modeling tool

The UML modeling tool Enterprise Architect (EA) was developed and released by the Australian company Sparx Systems. It is a CASE (Computer Aided Software Engineering) tool for designing and constructing software systems. The modeling tool supports all UML features defined in the UML 2.0 specification [UMb04]. The modeler can create all the 13 UML 2.0 diagrams by dragging and dropping predefined model elements onto a canvas in the program window. By switching to the UML 1.4 meta model, the modeler can use the stereotypes defined in this version as well. Furthermore Enterprise Architect offers a lot of useful functions for modeling with UML:

Properties of the Enterprise Architect UML tool

- Creating UML model elements for a wide range of purpose
- Placing those elements in diagrams and packages
- Creating connectors between elements
- Documenting these elements
- Generating code for software development
- Reverse engineering of existing code in different languages

We chose Enterprise Architect as the UML environment for the UMM Add-In because the software is inexpensive and a popular tool for modeling business models. It is used by a lot of small and medium-sized enterprises (SMEs) representing most of the present and future UMM users.

Enterprise Architect is suitable for adding the UMM Add-In

A further reason is the communication interface for adding software modules to the Enterprise Architect. The Automation Interface provides a way for accessing internal elements of the Enterprise Architect model. Furthermore the interface makes it easy to manipulate the contents of Enterprise Architect program components. Using methods of the interface, values of model elements in the UML model can be processed. In the UMM Add-In the Automation Interface was used for:

The Automation Interface of the Enterprise Architect

- Editing tagged values, diagrams and model elements
- Creating tagged values, diagrams and model elements
- Deleting tagged values and model elements

ActiveX COM compliant programming languages are able to connect to the Enterprise Architect Automation Interface. Programing languages satisfying these requirements are Microsoft Visual Basic 6.0, Borland Delphi 7.0 and the .NET Framework. The UMM Add-In is developed in C# using the Microsoft .Net Framework 2.0.

Using Microsoft C# for the UMM Add-In

4.3   Software development environment

As mentioned before the UMM Add-In is developed in C#. Microsoft Visual Studio .Net 2005 has been our tool of choice for the implementation. Using this development environment the automation interface needs to be integrated to the Common Object Model (COM) library of Microsoft Visual Studio. With this step all classes provided by the Enterprise Architect are imported to the environment. This guarantees full access to the methods and variables the UML tool uses for displaying and processing UML model elements.

Extending the COM library of Microsoft Visual Studio

For registering the COM objects of the Enterprise Architect the application needs to be registered in the Windows Registry. Once this step is done a properties window in the Microsoft .NET environment enables setting the references to other COM libraries. This properties window is depicted in Figure 4–1. The Common Object Model for the Enterprise Architect is called Enterprise Architect Object Model 2.10 and needs to be imported before implementing plug-ins for Enterprise Architect.

Setting the reference to the EA Object Model

Fig. 4–1 Adding the EA Object Model to Visual Studio .NET 2005

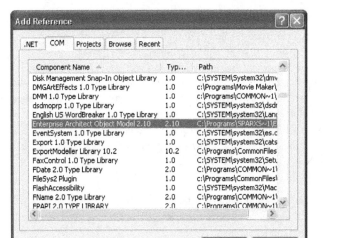

The following code listing shows an example of using methods of the Enterprise Architect Object Model. The example localizes a specific package in the UMM model and adds a new model element to this package. In codeline 4 a new package is instanced by retrieving the model by a specific ID. The next line adds a new subordinated model element to this package. This element is of type *use case* and is named *register customer*. In codeline

Using different methods of the EA Object Model

6 the UMM stereotype *business process* is assigned to this element. The *Update()* method in codeline 7 confirms the creation of the new element.

```
[4]   EA.Package Package = Repository.GetPackageByID(2);
[5]   EA.Element Element
      = (EA.Element)Package.Elements.AddNew("Register Customer", "UseCase");
[6]   Element.Stereotype = "BusinessProcess";
[7]   Element.Update();
```

Listing 4–1 Example for adding a new model element with the EA Object Model

As we can see, the components of an Enterprise Architect model can be accessed and altered very easily. Although other tools like Rational Rose offer similar APIs in order to access program components, the one provided by Enterprise Architect was better tailored to our requirements. The creation of the UMM Add-In are greatly enhanced, because the features of the Automation Interface was thoroughly documented. Finally we would like say, that the support provided by Sparx Systems is fast and state of the art.

5 UMM at a glance

5.1 About UN/CEFACT

The United Nations Economic Commission for Europe (UN/ECE) founded the Centre for Trade Facilitation and Electronic Business (UN/CEFACT) for improving the worldwide coordination of trade. UN/CEFACT is a long-existing B2B standards body, which became famous by developing and maintaining the UN/EDIFACT standards. In particular UN/CEFACT improves the ability of business, trade and administrative organizations, from developed, developing and transitional economies, to exchange products and relevant services effectively. The organization focuses on facilitating national and international transactions, through the simplification and harmonisation of processes, procedures and information flows. UN/CEFACT argues for the growth of global commerce [UNC05].

The task of UN/CEFACT

The organizational chart in Figure 5–1 shows the structure of the UN/CEFACT permanent working groups.

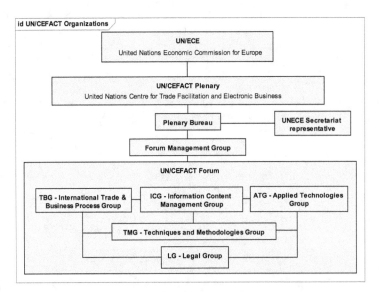

Fig. 5–1 Structure of the UN/CEFACT permanent working group [UNC05]

UN/CEFACT consists of five UN/CEFACT groups depicted in the bottom of Figure 5–1 are established to form the UN/CEFACT development structure. While TBG (International Trade and Business Processes Group), ICG (Information Content Management Group) and ATG (Applied Technologies Group) are the operational groups, the TMG (Techniques and Methodologies Group) and LG (Legal Group) are serving essentially as support groups [CSO02]. *The five UN/CEFACT Groups*

The TBG is responsible for business and governmental process analysis and international trade procedures using the UN/CEFACT Modeling Methodology. Thus it supports the development of appropriate trade facilitation and electronic business solutions, including the development and maintenance of UN and UN/ECE recommendations. The ICG is responsible for the management and definition of reusable information blocks retained in a series of libraries. The ATG is responsible for the creation of the trade, business and administration document structures that will be deployed by a specific technology or standard such as UN/EDIFACT or XML. The TMG is responsible for providing all UN/CEFACT groups with base ICT (Information and Communications Technology) specifications and recommendations. Furthermore the TMG is an ICT research group. This group is developing the next generation of EDI (Electronic Data Interchange) and is also responsible for the development of UN/CEFACT's Modeling Methodology (UMM). The Legal Group (LG) is responsible for publishing and presenting analysis and recommendations regarding legal matters related to UN/CEFACT. *The tasks of the five Permanent Working Groups*

UN/CEFACT and UMM

As we can see all these five permanent working groups have their own purposes and tasks. The TMG (Techniques and Methodologies Group) develops the UN/CEFACT Modeling Methodology (UMM) continuing the work of the former TMWG (Techniques and Methodology Working Group). The TMG produces trade facilitation, electronic business recommendations and technical specifications to advance global commerce. Thus an important task of the TMG is to provide a clear specification of UMM to familiarize business domain experts with this new modeling methodology. We joined several meetings of UN/CEFACT to contribute to the work of the TMG. We worked together with other TMG members on the definition and the documentation of the UMM meta model. *The TMG developed the UN/CEFACT Modeling Methodology (UMM)*

5.2   Basics of the UMM

Sometimes a business environment is large and complex. In order to understand the basics of this environment, we have to begin with collecting infor- *What is UMM?*

mation and documenting the business domain knowledge. The UMM is an incremental methodology for modeling business processes. It provides different views on interorganizational business processes suitable for communicating the model to business practitioners, business application integrators, and network application solution provider. Furthermore the UMM provides the conceptual framework to communicate common concepts. UN/CEFACT's Modeling Methodology sits on top of the Unified Modeling Language (UML).

UN/CEFACT has developed UMM for providing a methodology to capture business process knowledge, independent on the underlying implemented technology. UMM is based on the Open-edi reference model which concentrates on the business semantics of a B2B partnership. Open-edi describes two views – the *business operational view* (BOV) and the *functional service view* (FSV) [OER95]. The BOV describes the business processes in a format that is independent from any programming language. UN/CEFACT proposes UMM for modeling the BOV. The FSV describes the technical framework used to discover and transport the business information. Figure 5–2 shows these two views as the basic concept of the ISO Open-edi reference model. The goal of Open-edi is capturing the commitments made by business partners, which are reflected in the resulting choreography of the business collaboration. UMM is the formal methodology for describing an Open-edi scenario as defined in the Open-edi reference model. Moreover it is providing methods for specifying collaborative business processes involving information exchange in a technology-neutral and implementation-independent way.

What are the base concepts of UMM?

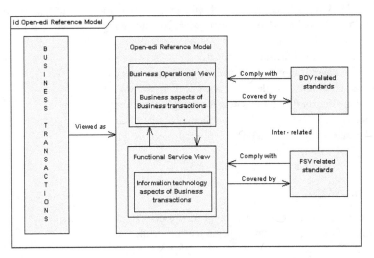

Fig. 5–2 The Open-edi Reference Model

UMM vs. UML

UMM is based on the concepts of the Unified Modeling Language (UML). All the UMM artifacts are documented in UML. More precisely UMM is defined as a UML Profile. The modeling methodology is based on a meta model, which is specified using the extension mechanisms of the UML. The UMM meta model defines a coherent set of *stereotypes*, *constraints*, and *tagged values*, i.e. a UML Profile for the purpose of modeling interorganizational business processes. The UMM meta model defines three different views represented as *packages* in the UMM model. In order to describe UMM compliant business collaboration models each view has its own semantics and stereotypes. These main views are listed as follows.

UMM uses UML for representing the model

- Business Domain View (BDV)
- Business Requirements View (BRV)
- Business Transaction View (BTV)

5.3 Business Modeling using the UMM

Building a UMM compliant business model requires a top-down modeling approach. It has to begin with a clear understanding of the business domain and the business activities therein. This approach requires the definition of business entities, their state management, and state life cycle identification to produce a model that can evolve as a new business requirements emerge. On the other side the bottom-up approach can be used as a starting point to incorporate existing and well-known business documents and transactions. Moreover this approach helps identifying some model elements and must be applied in order to produce evolvable and maintainable models that support reusing business processes between trading partners on the Internet [UG03].

Top-down approach vs. bottom-up approach

Business information dependencies, not document exchange

The task of UMM is to formalize dependencies between business partners for a business domain. Old approaches focused only on the document structure of documents exchanged. UMM instead focuses on business actions and objects that interact with business information.

UMM is not a document exchange methodology

Model production approach

The UMM uses worksheets to capture the business domain knowledge. These worksheets are simple tools to collect and organize the information needed to produce the minimum UMM models for each work area. There is an iterative process of gathering the information for the various work areas. The information collected in interviews with business stakeholders needs to be captured in the worksheets of the BDV, BRV and BTV.

Worksheets are capturing the business domain knowledge

5.3.1 Business Domain View

The approach taken in the *business domain view* (BDV) is an interview process between the modeler and the business domain expert. The purpose of this interview is to discover internal or interorganizational business processes. Furthermore the task of this view is to capture these business processes and to find business partners participating in these processes. These artifacts are represented as *use case diagrams* in the UMM model. The result of the business domain view should not be the construction of new business processes but discovering business processes and capturing their knowledge within worksheets. This view determines the business context of the process for finding reusable, previously defined, process descriptions or terminology in the UMM libraries.

The task of the business domain view

Furthermore the BDV is using existing knowledge. The business processes are classified according to a classification scheme. Candidate schemes are:

Using predefined classification schemes

- Porter's Value Chain
- SCOR (Supply Chain Operations Reference Model)
- UN/CEFACT's Common Business Processes Catalog

5.3.2 Business Requirements View

The *business requirements view* (BRV) builds up on the *business domain view*. This view captures the business scenarios, inputs, outputs, constraints and boundaries for business processes. It describes scenarios where the business partners in the domain under consideration can collaborate. The result of this view is a description how the business domain expert sees and describes the collaboration to be modelled. The BDV is expressed in the language and concepts of the business domain expert.

The BRV captures business scenarios of possible business collaborations

The definition of a business process is an organized group of related activities that together create customer value [HC93]. The business collaboration has a similar definition. It is a special kind of a business process, where the activities are executed by two or more business partners. The collaboration is called a binary collaboration in case of two business partners and multi-party collaboration in case of more than two business partners.

The definition of a business collaboration

The first step in the BRV is to get a common overview of potential business collaborations. *Activity graphs* are utilized to informally specify the flow of these collaborations. Figure 5–3 depicts steps of a collaboration dealing with a registration process between a *customer* and a *registrar*. Each business partner is represented by his own *partition*. The *shared business entity state* between the partitions denotes a need of an interaction between both partners' e-business systems.

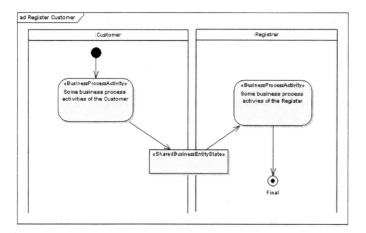

Fig. 5–3 Example process flow depicted by means of the BRV

Requirements and participating roles of a collaboration are formally specified using *business collaboration use cases*. Figure 5–4 shows the business collaboration use case *register customer* capturing the requirements for the process shown in Figure 5–3. The interaction between the two systems is manifested by the *business transaction use case*. *Customer* and *registrar* are specified as roles participating in the *register customer* process.

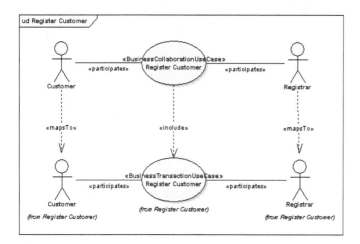

Fig. 5–4 Capturing the requirements of the register customer collaboration

5.3.3   Business Transaction View

The *business transaction view* (BTV) captures the semantics of busi- *The task of the BTV*
ness information entities and their flow of exchange between roles as they
perform business activities. The task of this view is to transform the requirements from the BRV into an analysis model. Thus the BTV uses the language and the concepts of the business analyst and describes how the business analyst sees the process to be modeled. There are three important
artifacts of the business transaction view:

- Activity graph for a business collaboration
- Activity graph for a business transaction
- Class diagram describing the data exchange

Taking a look at this list, we have to make a distinction between a busi- *Distinction between busi-*
ness collaboration and a *business transaction*. A business transaction is the *ness collaboration and*
basic concept for defining a choreography of a collaboration between busi- *business transaction*
ness partners. Communication in a business collaboration means that all relevant business objects are in the same state in each information system. If
this state is changing for any reason, a business transaction is initiated to
synchronize the states in both information systems.

The first bullet describes the activity graph for a business collaboration. *Activity graph for a busi-*
This activity graph specifies the choreography of activities among two or *ness collaboration*
more business partners. Furthermore the business collaboration is modeled
according to the requirements defined in the BRV. Figure 5–5 depicts our
example *register customer* collaboration. This activity graph is composed
of one *business transaction activity* denoting the only required interaction
between the two partners's systems.

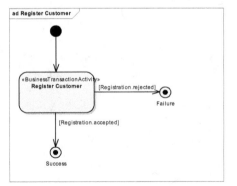

*Fig. 5–5 Activity graph
for the business
collaboration register
customer*

The second type of activity graph describes a business transaction in more detail. There are two different types of business transactions. The first type is called one-way business transaction and describes the following scenario: the requesting business partner reports a state change that the reacting business partner has to accept. In this case information is changed in only one direction. The responding business partner is receiving the request from the requesting business partner, but there is no channel for returning information to the initiating business partner. The second type of a business transaction is called two-way transaction. In this case the requesting business partner sets the information to an interim state. The requesting partner changes this state. Since there is a bidirectional connection between the two business partners the reacting business partner can define this state he already changed as the final state. There is an important relationship between business transaction use cases in the BRV and the business transactions in the BTV. The requirements (e.g. the participants) for modeling a transaction are defined in the BRV. Figure 5–6 shows a business transaction called *register customer*. This business transaction is the detailed description of the equally named business transaction activity which is part of the collaboration depicted in Figure 5–5. In this case the business transaction is a two-way transaction. Each partition in the activity graph is executed by a role of the corresponding use case in the BRV.

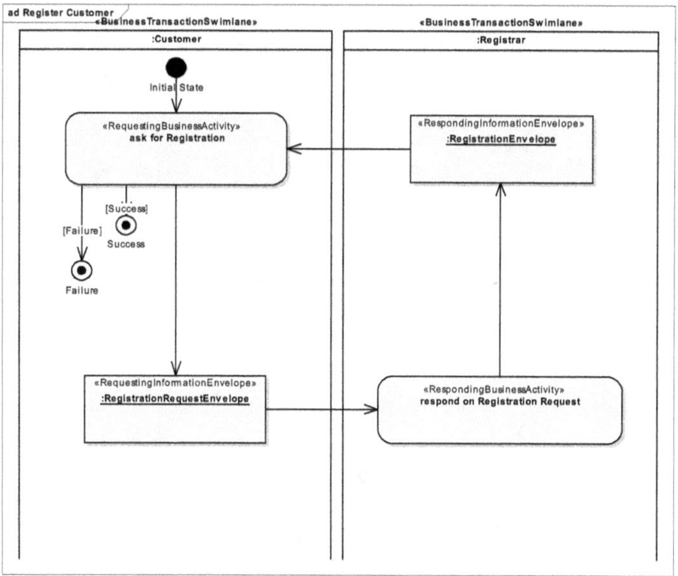

Activity graph for a business transaction

Fig. 5–6 Business transaction register customer

The third artifact of this view is the *class diagram* for describing the data exchange. In UMM it is also important to know everything about the information structure. This structure is composed of the business information entities included in each single business transaction. The business entities are described by the information needed to change its business states. Figure 5–7 depicts the structure of the *quote envelope*. Please note that this cutout has been taken from a different example. In UMM an *information envelope* is associated with one *information entity* defined as *header* and with at least one information entity defined as *body*. An information entity represents the actual business document.

Class diagram for describing the data exchange

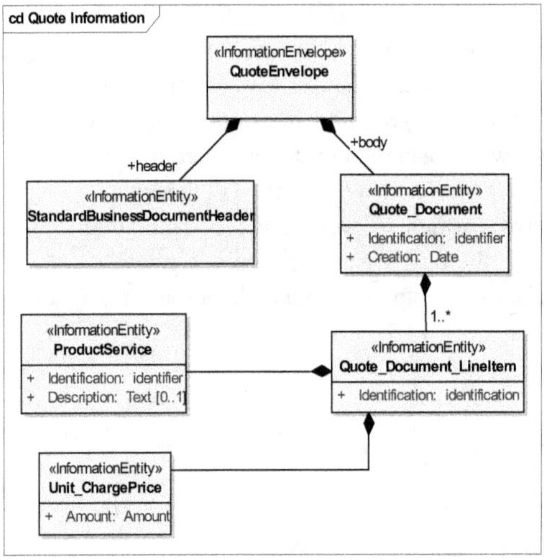

Fig. 5–7 Information structure

6   Worksheets Editor

6.1   The need for worksheets

Before the modeler is going to design a business model, he needs to get
an overview about the business domain knowledge. In UMM the business
process analyst gathers information that is important for the modeling process. Thus the modeler is able to redesign or refine business processes. In
UMM a set of worksheets helps him to capture this information. The UMM
worksheets are a set of documents capturing the business domain knowledge. These documents have their fixed predefined structure and have originally been defined in Microsoft Word format. In order to keep them consistent and readable for different people the documenting design-rules must
not change.

*Before to start modeling
with UMM*

Worksheets do not only help the designer to keep his own notes, but
provide a means of communicating the business model in natural language
to business people. In most cases business domain experts do not understand the meaning of modeling artifacts represented in a UMM diagram. It
is easier to understand, if there is a plain text, where almost the whole model
is described. Furthermore this is exactly the way of communication between
the businessman and the modeler. The use of modeling tools on the modeler's side and the use of text-based forms in another software on the business man's side makes the scenario of refining the model over time more
complicated. It follows that modeling the business collaboration and documenting business knowledge in worksheets should be provided by the same
software. Thus we integrated the worksheets into the modeling tool.

*Worksheets provide a
means of communication
between the modeler and
the businessmen*

Figure 6–1 shows an example of documenting a business area. This
model element represents a package in the business domain view to circum-
scribe the business areas. It is used to document, which kind of business
processes and artifacts are defined in the package named *procurement/sales*.

Fig. 6–1 Worksheet of a model element stereotyped as business area

Form: BusinessArea	
General	
Business Area	Procurement/Sales
Description	In this business area, business processes are described where a purchasing organisation can find potential suppliers for required products, can establish an account with the selling organisation, request for a quotation of required products and eventually place a purchase order with the selling organisation if the quote provided by the selling organisation meets the purchasing organisation's business objectives.
Objective	The objective of this business area allows a purchasing organisation to find an appropriate supplier (selling organisation), to establish an account, to request a quote for required products, and finally to purchase these products.
Scope	- Identify potential customer/ vendor - Request quote for price and availability - Request purchase order
Business Opportunity	The business opportunity of this business area is to allow purchasing organisations to purchase required products from selling organisations.
Business Library Information	
Base URN	http://www.untmg.org/UserGuide2005/BDV/Procurement
Version	0.1
Status	approved
Business Term	Purchase Order, Order, RFQ, Quote, Quotation, Sales Order, Price Request
Owner	UN/CEFACT
Copyright	UN/CEFACT
Reference(s)	

Fig. 6–1 Worksheet of a model element stereotyped as business area

6.2   The integration of worksheets into a UMM tool

The usual workflow using worksheets for documentation is as follows. The business analyst works together with the business domain expert in order to gather the business information that is of interest for creating a B2B collaboration model while going from one worksheet to another. After the business analyst has enough information, he is going to design the UMM model with an UML Tool. A big disadvantage of this workflow is that the modeler cannot check any eventual flaws in the model during the meeting with the business domain expert. These flaws can eventually force an inefficient model. Furthermore the approach requires additional meetings and communication-effort.

Workflow while using no integration of the worksheets to a tool

In order to improve the workflow, it is important to integrate the worksheets into the modeling tool. In this case the modeler can immediately enter the business domain knowledge into a user-friendly windows-form of the modeling tool. While entering this information he can use interactive functions of the worksheet editor. As shown in Figure 6–2 the worksheet editor has four main tasks:

Workflow with interactive worksheets integrated to the modeling tool

Representation of the business domain knowledge

- Pattern generation
- Persisting information
- Exporting information

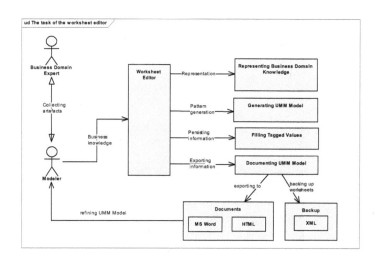

Fig. 6–2 The tasks of an interactive worksheet editor

The first task is to represent the business domain knowledge in a user-friendly manner to the modeler. In UMM different kind of worksheets have been designed. It follows that we also need different kind of windows forms. In other words each stereotype has its own distinct representation. Instead of having the business knowledge documented in a Microsoft Word document, an interactive data representation is more user-friendly and keeps the model consistent. Figure 6–3 shows an example of a windows form. The worksheet information displayed in this window represents the same information as the worksheet document in Figure 6–1. The worksheet of the stereotype *business area* is separated into different kinds of catego-ries—*General, Business Library Information, Process Area(s), Business Area(s)*. In the paper based worksheet in Figure 6–1 these categories are represented as sub-headings. In the user interface of the Figure 6–3 we use tabs to represent the categorization. Since there is a possibility to navigate through the structured information, the worksheet editor shows the data in a more customized and user-friendly way.

The representation of the worksheet data

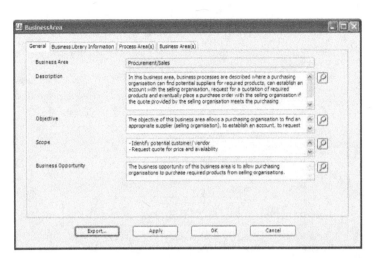

Fig. 6–3 Well structured
representation of the
worksheet data

Some UMM diagrams follow predefined patterns. Worksheets carry all
the information needed to apply a pattern for generating a diagram. In this
case we are able to automatically generate diagrams based on the business
knowledge captured in the worksheet editor (e.g. generating business transactions). This saves a lot of time by relieving the modeler from the routine
task of generating the same type of diagram over and over again. More
details about this tasks are described in chapter 6.7.1.

Generating UMM diagrams automatically with the worksheet editor

The next task is persisting the business domain knowledge. This feature
is fully integrated to the tool as well, because the information of the business knowledge is directly stored in the model. So-called tagged values
defined in the UML meta model are used to capture the worksheet information. The concept of the integration of the tagged values into the worksheet
editor is described in chapter 6.4.3.

Persisting business infor-mation

Furthermore these forms can be exported to the following output formats Microsoft Word, XML (WDL - Worksheet Definition Language) and
HTML. This last task is the communication interface between the business
domain expert and the modeler. With the documents generated out of these
output files, the UMM model is documented in a natural language, which is
readable for everyone. The way of defining the requirements through interactive worksheets makes this workflow more efficient and intuitive. More
details about the use of tagged values are described in the next chapter.

Exporting worksheets to Microsoft Word, HTML or XML

The next major advantage of the integration into a tool is the unique
evaluation. If the modeler designs a UMM model, some worksheet information may be required at different parts of the model. The business domain
knowledge captured in the beginning of the modeling process is required in

Unique evaluation of work-sheet entries

the end of the modeling process. For example a business transaction in the business transaction view is based on the business transaction use case in the business requirements view. If the modeler filled out the worksheet entries of the stereotype *business transaction use case*, this information is required in the assigned business transaction as well. Therefore the interactive worksheets will help him to keep the model consistent. Thus the modeler does not need to enter the same information twice.

6.3   Relationship between worksheets and tagged values

The worksheet information has to be stored somewhere. There are two alternatives to do this. The first one is to create a data file (e.g. in XML-format). Keeping the documentation - in our case the worksheet input - separated from the model itself can lead to inconsistencies.

Storing the information of the worksheets

Therefore the information needs to be saved in the model itself. The second alternative is the integration of the information into tagged values. The UMM meta model assigns tagged values to some significant model elements, which are comparable to attributes for storing further properties of this element. This is exactly the task of the worksheet editor – describing model elements. This step uses two advantages. On the one hand we integrated the worksheets into the model and on the other hand we can capture the mandatory tagged values specified by the UMM meta model.

Using tagged values to save worksheet entries

The second advantage needs to be inspected in more detail. The UMM meta model defines some tagged values which are assigned to specific model elements. If the modeler creates a clear and efficient model, these tagged values must be provided. Furthermore the modeler can add any amount of new tagged values to the predefined set. The worksheet editor integrates all tagged values predefined in the UMM meta model.

The tagged values pre-defined in the UMM meta model are filled out by the worksheets

Figure 6–4 shows the representation of the tagged values in the modeling tool Enterprise Architect. The UMM meta model defines the following tagged values for the stereotype *business area*.

objective
scope
businessOpportunity
baseURN
owner
copyright
reference
version
status
businessTerm

As seen in the Figure 6–4 all tagged values are represented. Even if these tagged values are not predefined in the UMM profile, the worksheet editor creates a new tagged value, if the corresponding text field has been filled out. In other words, if the worksheet finds an entry, which has the same tagged value name, it will overwrite the value of this tagged value. Otherwise a new value will be created to store the new information. These tagged values should be empty at the beginning of the modeling process and should not be used for other purposes. Furthermore the names of the tagged values must be unique, to avoid inconsistency.

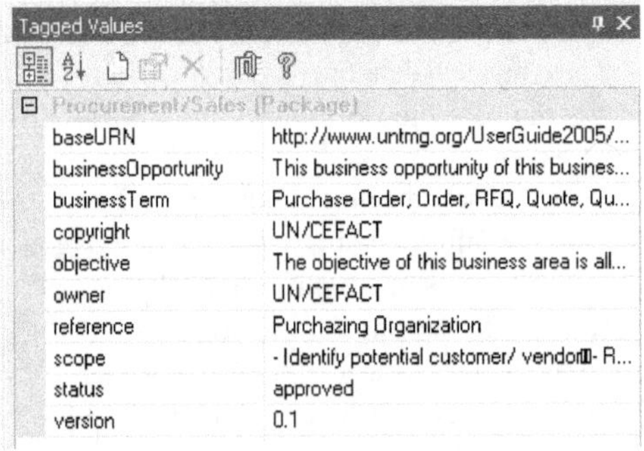

The worksheet editor creates new tagged values, if they do not exist

Fig. 6–4 The tagged values of a business area represented in the modeling tool

In addition to the predefined tagged values the modeler can add further ones. If he adds new tagged values, the worksheet editor is able to distinguish between customized tagged values and predefined tagged values. Such customized tagged values should also be editable via the worksheet editor. In this case, the worksheet editor also serves as a "tagged value editor". Such special tagged values are listed in a separate category.

Worksheet editor is also a tagged values editor'

The worksheet editor does not only store the business knowledge in tagged values. Some information is stored in UML constraints or internal variables of the modeling tool. Starting with the first option, we know, that UML offers predefined *constraints*. For example the information about the *pre- and post-condition* of an element are such *constraints* as already defined by the UML. The other option are internal variables of the modeling tool. For example the notes field that may be added to any UML element represents such a variable. More detailed information about these special ways of saving the information is provided in chapter 6.4.3.

Storing constraints and further information to the internal structure of the modeling tool

6.4   Technical implementation of the worksheet editor

One of the requirements was to design user-friendly windows forms which guide the modeler step by step to an efficient model. For this reason

Requirements of the worksheet editor

the worksheet editor provides input form windows with structured information on it. The text fields are separated by categories classifying the different types of business knowledge.

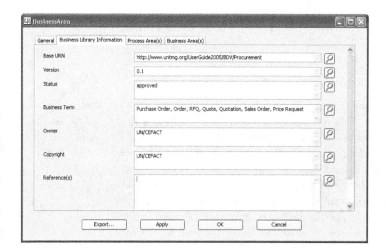

Fig. 6–5 The worksheet editor window of a business area

Figure 6–5 shows a screenshot of a worksheet editor window for a model element stereotyped as *business area*. The input fields represent the worksheet entries and each different tab separates the information into categories. The information shown in this figure is equivalent to the general part of the worksheet document in Figure 6–1. Furthermore a click on the magnifying glass will pop up a window with a larger text area allowing more comfortable editing. This is especially useful for long texts. The two extra tabs called *Business Area(s)* and *Process Area(s)*, which are not shown in Figure 6–1, are extensions of the worksheet editor. They give information about the included business areas and process areas. The purpose of these extensions is explained in the chapter 6.7. The buttons at the bottom of the worksheet editor window have the following functions.

The user interface of the worksheet editor

Export button
Pressing this button executes the export functionality to export the worksheet information into different formats.

Apply button
Saves the worksheet information to the tagged values of the assigned model element without closing the window. After pressing this button, editing of the worksheet may be continued.

OK button
Saves the worksheet information to the tagged value of the assigned model element and closes the window. The editing session of this worksheet is finished after pressing the OK button.

Cancel button
Pressing this button aborts the editing session without saving any information to the tagged values.

6.4.1   The need for a dynamical structure

Large scope projects spanning over a lot of business processes lead to many required artifacts before designing a model. This results in a large number of different worksheets for all different stereotypes. As a consequence we need to be able to maintain the worksheets quickly and easily. Thus it is not efficient to hard-code the worksheet layout in the C# code. Instead we define the layout of worksheets in special XML encoded data files. The UMM Add-In dynamically loads these data files and renders the worksheet according to its layout definition. This means that the maintenance of the worksheets does not require to change the code of the UMM Add-In. It is simply realized by updating the XML data file.

Using a dynamical structure for an easier maintenance

The diagram in Figure 6–6 shows the scenario of rendering of the windows form, based on the XML data files. Each stereotype which has an attached worksheet, needs its own XML data file. These XML files are described in Worksheet Definition Language (WDL) and describe the layout information of the specific worksheet. A WDL file for example defines the layout of each input field, the access state (read-only, write-protected) and the functionality of the input fields (drop-down box or text-area). WDL is described in more detail in the next sub-chapter.

Fig. 6–6 The design of the worksheet editor depends on the WDL input file

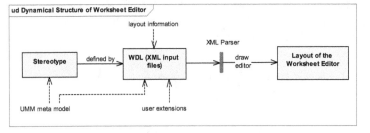

As we mentioned above, the maintenance plays a major role in the *If the UMM meta model* UMM Add-In. Since the modeling language is in a rather early stage and *changes, newly shaped* not yet mature, the meta model may change and improve over time. As seen *worksheets are required* in Figure 6–6 the stereotypes and their tagged values are defined in the UMM meta model. Since the worksheets must reflect the tagged values, the UMM meta model has a great impact on WDL. In case of a change in the meta model, the structure of the worksheets will change, too. Changing the code of the UMM Add-In every time is inefficient. So whenever the meta model is changing, the modeler just adds some tags to the WDL file and the worksheet window is adjusted accordingly. Furthermore the modeler may customize the worksheet editor while adding information to the input files. For example if he wants to have some additional fields, he can extend the functionality of the worksheets. This scenario is visualized as *user extensions* in the figure above.

The worksheet editor provides a feature to export the worksheets to dif- *Reusing the WDL output* ferent formats. One of these formats is WDL. Using this feature the modeler *files as input files* can create his own input file for the worksheet editor, because the file he exported is valid against the same WDL schema used from the standard input files. For example if he is working on the worksheet for the stereotype *business process* and he is going to create an WDL export file, including all entries he already made, he can reuse it as an WDL input file for another business process, which may be quite similar to the last one. In this case he saves time by not entering almost the same information twice. This scenario is visualized in Figure 6–7. The worksheet editor uses the WDL input file to receive instructions for the style of each text field in the editor. This WDL file is the definition file for the style of any model element stereotyped as *business process*. After the modeler added the worksheet information for the business process *provide product list*, he wants to edit another business process called *update product list*. These processes are similar and thus their worksheets are similar as well. Therefore the modeler can export the worksheet of the first business process into the WDL format while using the export functionality of the worksheet editor. When the modeler begins the worksheet editing session for the next business process he can reuse the exported WDL file from the business process *provide product list*. Since the data has been exported as well, the worksheet of the business process *update product list*, is initialized with the exported data.

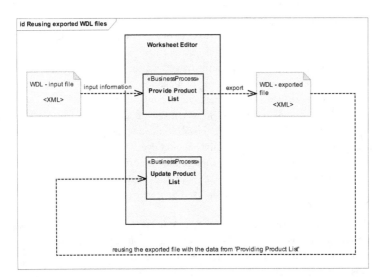

*Fig. 6–7 Reusing
worksheet information
from another model
element*

6.4.2   WDL - Worksheet Definition Language

In order to provide high flexibility, WDL is used to define the style of any worksheet editor window. Figure 6–8 depicts the technical implementation of displaying a worksheet form for the stereotype *business area*. As mentioned before, different types of stereotypes need different types of worksheets. Consequently stereotypes build the basis for the selection of the worksheets. For example, if the modeler chooses a model element in Enterprise Architect stereotyped as *business area* with the intention to edit this worksheet, there is a WDL file telling the UMM Add-In to use the *business area* input file. This "deployment file" is called *DefaultWorkSheets.xml*. So whenever the name of a stereotype is changing (e.g. as the consequence of a change or an extension of the UMM meta model) it needs to be documented in this file.

WDL defines the style of the worksheets

After the corresponding stereotype was identified and assigned to a WDL input file, a method checks if the input file exists. If the file *BusinessArea.xml* was found, its content is parsed. The XML parser processes the XML instance and displays the worksheet as it was defined in the file. Eventually some extensions for displaying singularities of specific stereotypes may be added. The stereotype *business area* for example needs such an extension for displaying all subordinated packages stereotyped as *business areas* and *process areas*. Since the WDL input file offers only static facts predefined by the UMM meta model and extensions defined by the

From WDL to worksheets

modeler himself, worksheet extensions integrate dynamical information depending on the model. As shown in Figure 6–5 the worksheet has two additional tabs for editing the business areas and the process areas.

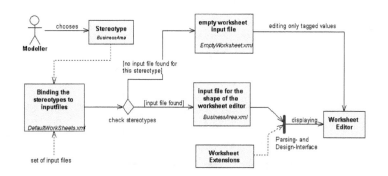

Fig. 6–8 Using WDL for designing the style of a worksheet editor window

In some cases there are stereotypes, which are not assigned to a worksheet. This could happen, if it is not necessary to document properties about the specific model element. However the modeler should be able to edit the tagged values of the stereotype. Thus there is a possibility to manage only the tagged values with the use of the WDL input file *EmptyWorksheet.xml*. This input file gives the instructions, that there are only two tabs shown in the editor window. One of them shows information about the model element and the other offers the input forms for editing the tagged values. If the deployment file binds the stereotype to the right WDL input file and for any reason the specific input file is not found, the *EmptyWorksheet.xml* file is used instead. Figure 6–9 shows the editor window of this scenario. The stereotype *business partner* does not require his own worksheet, but there is a tagged value specified by the UMM meta model called *interest*. Thus the editor provides a feature for editing such a non-categorized tagged value.

Some worksheets are not assigned to a stereotype

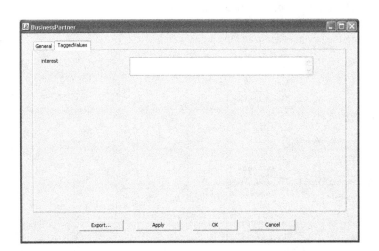

Fig. 6–9 Editing a
worksheet which is not
assigned to a stereotype

6.4.2.1   The W3C schema of WDL

The schema of the input files is designed to be easy to understand, because modelers should be able to create their own worksheet definition files. The following code shows an excerpt of the XML schema definition of WDL. Before a new worksheet is loaded, the UMM Add-In checks if the WDL file for this worksheet is wellformed and valid against this schema. In case of an invalid WDL document, an error message appears. This prevents the application crashing due to malfunctioned worksheet definitions.

WDL validation with the
before displaying the work-
sheet editor window

```
[8]    <?xml version="1.0" encoding="UTF-8"?>
[9]    <xs:schema xmlns:xs="http://www.w3.org/2001/XMLSchema"
           elementFormDefault="qualified" attributeFormDefault="unqualified">
[10]       <!-- === ROOT element === -->
[11]       <xs:element name="WORKSHEET" type="worksheetType"/>
[12]       <!-- === Element definitions === -->
[13]       <xs:element name="CATEGORY" type="categoryType"/>
[14]       <xs:element name="CHOICEBOX" type="choiceboxType"/>
[15]       <xs:element name="DEFAULT" type="xs:string"/>
[16]       <xs:element name="DEFINITION" type="xs:string"/>
[17]       <xs:element name="ENTRY" type="entryType"/>
[18]       <xs:element name="ITEM" type="xs:string"/>
[19]       <xs:element name="NAME" type="xs:string"/>
[20]       <xs:element name="SUBMENU" type="submenuType"/>
[21]       <xs:element name="TOOLTIP" type="xs:string"/>
[22]       <!-- === Attribute definitions === -->
[23]       <xs:attribute name="id" type="xs:ID"/>
```

Listing 6–1 The Schema
of WDL

```
[24]        <xs:attribute name="lines" type="linesType"/>
[25]        <xs:attribute name="name" type="xs:string"/>
[26]        <xs:attribute name="protected" type="xs:boolean"/>
[27]        <xs:attribute name="selected" type="xs:boolean"/>
[28]        <xs:attribute name="submenu" type="xs:IDREF"/>
[29]        <xs:attribute name="taggedValueName" type="xs:string"/>
[30]        <xs:attribute name="taggedValueType" type="taggedValueTypeType"/>
[31]        <xs:attribute name="type" type="typeType"/>
[32]        <!-- === Simple Types === -->
[33]        <xs:simpleType name="linesType">
[34]            <xs:restriction base="xs:integer">
[35]                <xs:minInclusive value="1"/>
[36]            </xs:restriction>
[37]        </xs:simpleType>
[38]        <xs:simpleType name="taggedValueTypeType">
[39]            <xs:restriction base="xs:string">
[40]                <xs:enumeration value="name"/>
[41]                <xs:enumeration value="notes"/>
[42]                <xs:enumeration value="constraint"/>
[43]                <xs:enumeration value="standard"/>
[44]                <xs:enumeration value="BusinessTransactionUseCase"/>
[45]                <xs:enumeration value="RequestingRole"/>
[46]                <xs:enumeration value="RequestingBusinessActivity"/>
[47]                <xs:enumeration value="RespondingRole"/>
[48]                <xs:enumeration value="RespondingBusinessActivity"/>
[49]                <xs:enumeration value="RequestingInformationEnvelope"/>
[50]                <xs:enumeration value="RespondingInformationEnvelope"/>
[51]            </xs:restriction>
[52]        </xs:simpleType>
[53]        <xs:simpleType name="typeType">
[54]            <xs:restriction base="xs:string">
[55]                <xs:enumeration value="text"/>
[56]                <xs:enumeration value="choice"/>
[57]                <xs:enumeration value="time"/>
[58]            </xs:restriction>
[59]        </xs:simpleType>
[60]        <!-- === Complex Types === -->
[61]        <xs:complexType name="categoryType">
[62]            <xs:sequence>
[63]                <xs:element ref="ENTRY" maxOccurs="unbounded"/>
[64]                <xs:element ref="SUBMENU" minOccurs="0"
                        maxOccurs="unbounded"/>
[65]            </xs:sequence>
[66]            <xs:attribute ref="name" use="required"/>
[67]        </xs:complexType>
[68]        <xs:complexType name="choiceboxType">
[69]            <xs:sequence>
[70]                <xs:element ref="ITEM" maxOccurs="unbounded"/>
[71]            </xs:sequence>
```

```
[72]           <xs:attribute ref="selected" use="optional"/>
[73]       </xs:complexType>
[74]       <xs:complexType name="entryType">
[75]           <xs:sequence>
[76]               <xs:element ref="NAME"/>
[77]               <xs:element ref="DEFAULT" minOccurs="0"/>
[78]               <xs:element ref="TOOLTIP"/>
[79]               <xs:element ref="CHOICEBOX" minOccurs="0"/>
[80]           </xs:sequence>
[81]           <xs:attribute ref="type" use="required"/>
[82]           <xs:attribute ref="lines" use="required"/>
[83]           <xs:attribute ref="protected" use="required"/>
[84]           <xs:attribute ref="taggedValueName" use="required"/>
[85]           <xs:attribute ref="taggedValueType" use="required"/>
[86]           <xs:attribute ref="submenu" use="optional"/>
[87]       </xs:complexType>
[88]       <xs:complexType name="submenuType">
[89]           <xs:simpleContent>
[90]               <xs:extension base="xs:string">
[91]                   <xs:attribute ref="id"/>
[92]               </xs:extension>
[93]           </xs:simpleContent>
[94]       </xs:complexType>
[95]       <xs:complexType name="worksheetType">
[96]           <xs:sequence>
[97]               <xs:element ref="DEFINITION"/>
[98]               <xs:element ref="CATEGORY" maxOccurs="unbounded"/>
[99]           </xs:sequence>
[100]      </xs:complexType>
[101] </xs:schema>
```

The schema is separated into five logical parts. Between line 10 and *The logical parts of the* line 12 is the definition of the root element, between line 12 and line 22 the *schema* elements are defined, between line 22 and line 32 is the definition of the attributes, between line 32 and line 60 is the definition of the simple types and between line 60 and line 101 the complex types are described.

The root element - WORKSHEET
The element described in line 11 is called *WORKSHEET* which must be the *The WORKSHEET element* root element in every WDL instance. The type of this element is called *worksheetType* and is defined in line 95. This complex type has at least two child elements. The sequence defines, that there must be exactly one element called *DEFINITION* followed by at least one element called *CATE-GORY* under the root element.

CATEGORY
This element is defined in line 13 and is typed as *categoryType*. Instances of *The CATEGORY element* this type structure the worksheet information into logical parts. They must

have at least one element called *ENTRY* and an optional number of elements called *SUBMENU*. These corresponding multiplicities are defined in the sequence between line 62 and line 65.

ENTRY

The element definition in line 17 sets the type of an entry element. The *entryType* is specified in line 74 and is the most important element type. It defines the style and content of any worksheet input field. Between line 75 and line 80 the child elements of *ENTRY* are defined. There must be exactly one *NAME* element, an optional element called *DEFAULT*, exactly one element called *TOOLTIP* and an optional number of *CHOICEBOX* elements.

Setting the properties of the worksheet input field

The *ENTRY* element defines the following attributes: the attribute *type* specifies the type of the worksheet entry and is defined in line 81. This attribute is mandatory and specifies the type of the worksheet entry. The simple type specification is between line 53 and line 59. There is a set of allowed values defined. The following list describes the meanings of the possible values.

text
: If the attribute is specified by this value, the type of the entry value is a *string*. In this case the input form for this worksheet entry is a simple text box.

choice
: If the attribute is specified by this value, the input form of this worksheet entry is represented as a drop-down box. This value is used, if there is a predefined set of values. The modeler can choose a value of the drop-down list. The list of values represented in this drop-down box is defined by the element *CHOICEBOX*. Within this element, each list-item is marked by the element *ITEM*, which is defined in the schema in line 18.

time
: Many characteristics of a business process relate to timing constraints, e.g. maximum time to perform. The representation of days, hours, minutes and seconds is combined in a single *string*, where all this information is encoded. If the attribute is specified by the value *time*, the input form of this worksheet entry is represented as 4 text boxes. Each text box is requesting either a number for the amount of days or a number for the amount of hours or a number for the amount of minutes or a number for the amount of seconds. When stored into the tagged value in Enterprise Architect the encoded String could for example be "PT3H2M1S", which means 3 hours, 2 minutes and 1 second.

Figure 6–10 shows three input fields each using a different type.

The attribute in line 82 is called *line* and sets the number of lines of an input form. The attribute is mandatory as well. The next obligatory attribute is described in line 83 and is called *protected*. The definition of this attribute is specified in line 27 and is of type *boolean*. Thus this element can only be set to *true* or *false*. If this attribute is set to *true* the input field is in a write-protected state and cannot be edited by the modeler. In line 84 a further attribute of the *ENTRY* element is specified. It is called *taggedValueName* and defines the name of the tagged value. The value defined by this attribute is the unique key for a specific worksheet entry in the list of all tagged values. This attribute is obligatory as well.

Defining the size of the input field

The following attribute called *taggedValueType* is an important attribute. It is responsible for specifying the right type of the tagged value. Since there are different possibilities to save the value of a worksheet entry, this attribute specifies the location where to save the information. Thus this attribute is obligatory. Between line 38 and line 52 the simple type definition of this attribute defines a set of values the attribute can have. In the following list the meaning of these predefined values is explained. 1.) to 4.) give information about the way of saving the value of the worksheet entry. 5.) to 11.) give information about the location of the model elements where to store the values.

The attribute taggedValue-Type specifies the type of the tagged value

1. taggedValueType="name"
2. taggedValueType="notes"
3. taggedValueType="constraint"
4. taggedValueType="standard"
5. taggedValueType="BusinessTransactionUseCase"
6. taggedValueType="RequestingRoles"

7. taggedValueType="RequestingBusinessActivity"
8. taggedValueType="RespondingRole"
9. taggedValueType="RespondingBusinessActivity"
10. taggedValueType="RequestingInformationEnvelope"
11. taggedValueType="RespondingInformationEnvelope"

name

If the attribute is set to *name*, the value of the input field will not be stored as a tagged value. The meaning of this value is that the input field displays the name of the model element. The name of an element needs not to be saved in a tagged value.

notes

If the attribute is set to *notes*, the value of the input field is stored into the notes field of the model element. This field is represented as a text area and is provided by the modeling tool.

constraint

If the attribute is set to *constraint*, the value of a tagged value is stored into the UML constraints. Such constraints are represented in their own modeling tool windows. For further information about *constraints* see chapter 6.5.

standard

This means that the value of the input field is a standard entry, which has to be saved as a tagged value.

BusinessTransactionUseCase

If the selected model element links to another model element, which is stereotyped as a *business transaction use case*, the worksheet editor retrieves all the information of the related model element and displays its values. This scenario e.g. takes effect in storing and representing worksheet information of the business transaction. The worksheet of the business transaction gets information of the corresponding model element stereotyped as *business transaction use case*. After editing the worksheet, the new information will be stored in the model element where the information originally came from.

RequestingRole

This attribute indicates that the information relates to the requesting role in the worksheet for the *business transaction*. If this value is set, the worksheet entry will be stored as a tagged value in the requesting *business transaction swimlane* of the business transaction.

RequestingBusinessActivity
If the attribute in a business transaction is specified by this value, the worksheet information will be stored in the corresponding model element stereotyped as a *requesting business activity*.

RespondingRole
This attribute indicates that the information relates to the responding role in the worksheet for the *business transaction*. If this value is set, the worksheet entry will be stored as a tagged value in the corresponding model elements of the responding *business transaction swimlane*.

RespondingBusinessActivity
If the attribute in a business transaction is specified by this value, the worksheet information will be stored in the corresponding model element stereotyped as a *responding business activity*.

RequestingInformationEnvelope
If this value is set in a business transaction, the worksheet information will be stored in as a tagged value in the corresponding model element stereotyped as *requesting information envelope*.

RespondingInformationEnvelope
If this value is set in a business transaction, the worksheet information will be stored as a tagged value in the corresponding model element stereotyped as *requesting information envelope*.

The last possible attribute of the element *ENTRY* is called *submenu*. This attribute is defined in line 86 and can be used optionally. If it is included in the corresponding element, the worksheet editor structures all worksheet entries having the same attribute value within a subordinated box. As we can see in line 28 the attribute is typed as *IDREF*. This means that the value of this attribute is a reference to the definition of the header of the subordinated box. This definition is specified by using the element *SUBMENU*. The complex type of this element is defined in line 88 and captures an attribute called *ref* which is typed as *IDREF*. This attribute is the key for the name of the subordinated box. The following code lines show an example of a submenu.

Creating a subordinated box by using the element SUBMENU

```
[102] <CATEGORY name="Business Information Envelopes">
[103]     <ENTRY type="text" lines="2" submenu="sub1"
              protected="true" taggedValueName="InformationName"
              taggedValueType="RequestingInformationEnvelope">
[104]         <NAME>Information Name</NAME>
[105]         <TOOLTIP>Information Type</TOOLTIP>
[106]     </ENTRY>
[107]     <ENTRY type="text" lines="2" submenu="sub1"
              protected="false" taggedValueName="InformationState"
              taggedValueType="RequestingInformationEnvelope">
[108]         <NAME>Information State</NAME>
[109]         <TOOLTIP></TOOLTIP>
[110]     </ENTRY>
[111]     <ENTRY type="text" lines="2" submenu="sub2"
              protected="true" taggedValueName="InformationName"
              taggedValueType="RespondingInformationEnvelope">
[112]         <NAME>Information Name</NAME>
[113]         <TOOLTIP>Information Type</TOOLTIP>
[114]     </ENTRY>
[115]     <ENTRY type="text" lines="2" submenu="sub2"
              protected="false" taggedValueName="InformationState"
              taggedValueType="RespondingInformationEnvelope">
[116]         <NAME>Information State</NAME>
[117]         <TOOLTIP></TOOLTIP>
[118]     </ENTRY>
[119]     <SUBMENU id="sub1">Information Envelope from Requesting
              Business Activity</SUBMENU>
[120]     <SUBMENU id="sub2">Information Envelope from Responding
              Business Activity</SUBMENU>
[121] </CATEGORY>
```

Listing 6–2 Example of a submenu in the worksheet for a business transaction

Line 102 specifies a new category named *business information envelopes*, which is represented as a new tab. This category has 2 different sections. The first one is called *Information Envelope from Requesting Business Activity* and the second one is called *Information Envelope from Responding Business Activity.* These two sub-menus are defined in line 119 and line 120. Both *SUBMENU* elements are defining their own keys. In total there are 4 worksheet entries displayed in this tab. Two of them are shown in the first box and two of them are shown in the second box. E.g. the attribute called *submenu*, which is set to *sub1* in line 103 specifies, that the first worksheet entry should be displayed in the first box. The same attribute specified in line 111 e.g. references the second *SUBMENU* element. Thus this worksheet entry is displayed in the second box. Figure 6–11 shows the representation of the WDL code in the worksheet editor.

Explanation of the example for creating subordinated boxes in the worksheet editor

The schema file has to be in the XML folder of the UMM Add-In and has a write-protected status. Moreover a file is needed with the information which WDL file should be assigned to which stereotype. As mentioned before, different types of stereotypes need different types of worksheets. This is an important proposition, because the stereotypes form the basis for the selection of the worksheets. For example, if the modeler clicks a model element in the Enterprise Architect stereotyped as *business transaction use case* with the intention to edit the worksheet, this XML file tells the UMM Add-In to take the *business transaction use case* input file.

The need for a deployment file for binding stereotypes to WDL input files

Fig. 6–11 Structuring worksheet entries

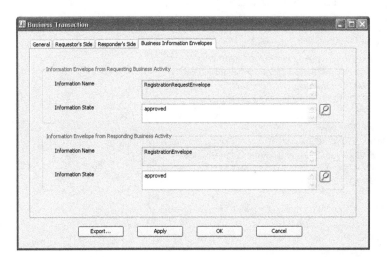

6.4.2.1 The WDL instances

Every stereotype, which is assigned to a worksheet by the UMM meta model has its own WDL instance. The final product of the worksheet editor is a set of documents with the information about the business domain knowledge captured in a UMM model. The structure of such a document is shown in Figure 6–1. As we can see, different sections may be displayed grouping entries logically belonging together. Furthermore the modeler is able to add subcategories. All these requirements of the output file are considered in the WDL instances. Therefore the structure of the input files is quite similar to the structure of the *Microsoft Word* output files or the *HTML* file generated after editing a worksheet.

The structure of the WDL input files is similar to the structure of worksheet documents

Before explaining all attributes and tags, we summarize the requirements for the design of the worksheet editor. Figure 6–12 shows all the factors influencing the design of the worksheet editor window. The bubbles around the rectangle represent these factors. They define how every single

Requirements for the design of the worksheet editor

worksheet has to look like and which special features the worksheet requires.

The factor named *categories* visualizes the possibility to separate the worksheet form into different logical parts. This feature increases the usability on the modeler's side and the readability on the business domain expert's side after exporting the worksheet. Each category captures issues belonging together. In the worksheet of Figure 6–1 there are two different categories. The first category is named *General* and the second one is named *Business Library Information*. Thus it is easy to read this structured content within the document and even more easy to distinguish logical parts of the worksheet editor while entering the business domain knowledge.

Providing categories to increase usability

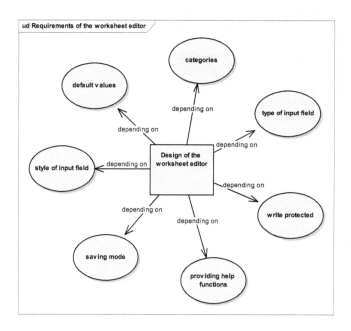

Fig. 6–12 Factors influencing the style of the worksheet editor

The next significant factor is the style of the input fields. This require-ment has the most attributes, because there are different ways and settings for adjusting the style of the input fields of the worksheet editor. For exam-ple the size of the input field is an important property. Sometimes business information is short and in many cases just one word, but sometimes busi-ness artifacts are described in a couple of paragraphs using a lot of space for its text. Thus a feature is required to calibrate the size (number of lines) of an input field.

Customizing the style of the input fields

Furthermore the functionality of an input field depends on the settings of an attribute named *type*. In some cases the worksheet editor should offer the modeler predefined values of a specific worksheet entry. These values are represented in a drop-down box. If free text is required, the attribute *type* must be set to *text*. Any input field may be initialized with a predefined default value. It does not matter if it is a text field or a drop-down box. Sometimes it is required to deny editing an entry value. This case occurs when the specific entry references values of another stereotype. If the *protected* attribute is set to *true* the modeler is not able to change the state of the input field. Thus the value is write protected and cannot be changed.

The attribute type is responsible for the style of the input fields

Because there are a lot of different worksheets, the modeler must be supported by help functions during the evaluation workflow. A help text can be added to any worksheet entry in form of a *tooltip*-helpbox. After setting the *TOOLTIP* tag to a specific text, a box representing this text will appear if the modeler moves the mouse pointer over the text field.

Help functions to support the modeler

The saving mode defines how the business domain knowledge is stored. Since there are different types of business information, there are also different ways of saving this information.

The attribute *taggedValueType* specifies the type of the tagged value. All allowed values of this attribute are in the schema definition for the worksheet input files between line 39 and line 51. These specific values initiate specific saving methods. For example, if the attribute is set to *standard*, the value of the input field is saved as a tagged value. If the attribute is set to *constraint*, the value will be saved as a UML constraint. In order to use the notes window in the modeling tool, the attribute's value must be *notes*. Another alternative is the name of another stereotype. In this case, the value of this attribute serves as a reference to a specific stereotype. This referenced stereotype provides the business information of the corresponding model element for initializing the input field. Thus this feature builds a bridge between different worksheets, because business information of other worksheets can be used by setting such references.

The attribute taggedValueType specifies the type of a tagged value

The following code shows an excerpt of an instance of a model element stereotyped as *business transaction*.

```
[122] <WORKSHEET xmlns:xsi="http://www.w3.org/2001/XMLSchema-instance"
           xsi:noNamespaceSchemaLocation="../worksheetSchema.xsd">
[123]     <DEFINITION>Name_of_EA_Stereotype</DEFINITION>
[124]     <CATEGORY name="General">
[125]         <ENTRY type="text" lines="1" protected="true"
                   taggedValueName="Name" taggedValueType="name">
[126]             <NAME>Business Transaction Name</NAME>
[127]             <DEFAULT>Name_Of_EA_Element</DEFAULT>
```

Listing 6–3 Excerpt of an WDL input file for a business transaction

```
[128]              <TOOLTIP>Provide a name for the business
                       transaction</TOOLTIP>
[129]          </ENTRY>
[130]          <ENTRY type="text" lines="2" protected="true"
                   taggedValueName="definition"
                   taggedValueType="BusinessTransactionUseCase">
[131]              <NAME>Definition</NAME>
[132]              <TOOLTIP></TOOLTIP>
[133]          </ENTRY>
[134]          <ENTRY type="text" lines="2" protected="true"
                   taggedValueName="purpose"
                   taggedValueType="BusinessTransactionUseCase">
[135]              <NAME>Purpose</NAME>
[136]              <TOOLTIP></TOOLTIP>
[137]          </ENTRY>
[138]          <ENTRY type="text" lines="4" protected="true"
                   taggedValueName="notes"
                   taggedValueType="BusinessTransactionUseCase">
[139]              <NAME>Description</NAME>
[140]              <TOOLTIP>A plain text explination of the purpose and
                       behavior of the Business Transaction.</TOOLTIP>
[141]          </ENTRY>
[142]          <ENTRY type="choice" lines="1" protected="false"
                   taggedValueName="BusinessTransactionPattern"
                   taggedValueType="standard">
[143]              <NAME>Select Business Transaction Pattern</NAME>
[144]              <TOOLTIP>Select one of</TOOLTIP>
[145]              <CHOICEBOX>
[146]                  <ITEM selected="true">Commercial Transaction</ITEM>
[147]                  <ITEM>Request Confirm</ITEM>
[148]                  <ITEM>Request Response</ITEM>
[149]                  <ITEM>Query Response</ITEM>
[150]                  <ITEM>Information Distribution</ITEM>
[151]                  <ITEM>Notification</ITEM>
[152]              </CHOICEBOX>
[153]          </ENTRY>
[154]          <ENTRY type="choice" lines="1" protected="false"
                   taggedValueName="isSecureTransportRequired"
                   taggedValueType="standard">
[155]              <NAME>Secure Transport</NAME>
[156]              <TOOLTIP>Select one of</TOOLTIP>
[157]              <CHOICEBOX>
[158]                  <ITEM selected="true">true</ITEM>
[159]                  <ITEM>false</ITEM>
[160]              </CHOICEBOX>
[161]          </ENTRY>
[162]      </CATEGORY>
[163] <!--     [...]          -->
```

[164] </WORKSHEET>

The root element in line 122 is called *WORKSHEET* and defines the location of the schema file. The root element has two child elements. The first one is called *DEFINITION* and keeps the header information for the worksheet. The content of this element is shown in the title bar of the worksheet editor. The expression in line 123 shows a special case of defining this header. The UMM Add-In provides predefined variables. Whenever these placeholders occur in the WDL input file, a parser is replacing them with the requested information of the model element. The variables used in the WDL input files are listed and explained in Table 6–4. This concept allows the modeler to add some specific information dynamically - e.g. the name of the element. In the WDL example above we use the variable *Name_of_EA_Stereotype*. This variable displays the stereotype of the model element. The same principle is used in line 127 where the variable *Name_of_EA_Element* is used for initializing the worksheet entry with the name of the modeling element.

Using variables for displaying properties of the model element

Tab. 6–4The meaning of the variables used in the WDL input file

Name of the variable	Description
Name_of_EA_Element	Returns the name of the model element, which is selected in the modeling tool.
Name_of_EA_Type	Returns the type of the model element, which is selected in the modeling tool.
Name_of_EA_Stereotype	Returns the stereotype of the model element, which is selected in the modeling tool.

The element *CATEGORY* in line 124 specifies a new worksheet category. The realization of this concept is represented as a new tab in the worksheet editor. The only attribute of this element is called *name* and specifies the name of this category, which is displayed in the header of the tab.

The element CATEGORY

Each definition of categories must have at least one *ENTRY* element defining the style of each input form of the worksheet editor. The element in line 125 includes a couple of attributes setting the properties of the worksheet entries. Since the attribute *type* is specified by the value *text*, the modeler may only add a simple String. Furthermore the attribute *lines* define the size of the input form. Different entries have a different character length. Thus the height of the input box is defined by this integer value. Since the

The ENTRY element and its attributes

name of the model element does not require so much space the size is *1*. Thus the input form offers only one line for entering the text. Line 125 specifies another attribute called *protected*. This attribute is set to *true* which means that the input field is write-protected. In this special case the worksheet entry is initialized by the name of the model element. Since this information is supported by the modeling tool and this worksheet entry does not need to be stored elsewhere, the modeler is prohibited of editing this entry. The attribute *taggedValueName* specifies the name of the tagged value. The value of this attribute will be added to the list of tagged values in the modeling tool, after the modeler gave the instructions for storing this worksheet entry. Furthermore the attribute *taggedValueType* is specified by the value *name*. In principal this attribute sets the type of the saving mode. The value *name* signalizes that this worksheet information does not need to be stored as a tagged value.

The element in line 126 is called *NAME* and defines the label of the worksheet entry. This label is displayed in the worksheet editor in the left column next to the input form. The following element in line 127 is called *DEFAULT* and initializes the input box with the value specified within this tag. As said before, the business transaction requests the information of the modeling element by variables. The content of the element called *TOOLTIP* is a helpful feature for supporting the modeler with help text. The content of this tag appears as a tooltip-feature, while moving the mouse over the specific input field or label. Thereby the modeler gets some instructions to complete the input fields. This help text is displayed in a yellow box, which disappears after a few seconds. This feature is implemented in line 128.

Initializing the worksheet entry with a specific value - the DEFAULT element.

In total, there are 6 worksheet entries described for this category in the WDL input file. The second one is specified in line 130 and has a write-protected state. The reason is the reference to another model element specified in the attribute *taggedValueType*. The business transaction is the implementation of a business transaction use case. Thus information is required from this use case. To create a relation between these two model elements, the *taggedValueType* must be specified by the value *business transaction use case*. This means that the initial value of this worksheet entry refers to a value of the corresponding business transaction use case. The only restriction is that the name of the tagged value of the worksheet entry must be the same as the tagged value of the related model element. E.g. in line 134 the tagged value called *purpose* of the corresponding business transaction use case is required. Therefore the input box is requesting the content of this tagged value with the same name to initialize the input field.

Getting worksheet information from other model elements

Line 142 outlines a special case. The attribute *type* is of value *choice*. This means that the worksheet editor displays a drop-down box instead of a simple text box. Since the business transaction can be described by 6 differ-

Using a drop-down list for offering a list of patterns

ent types of business transaction patterns, these patterns must be offered in a list. In line 145 the element called *CHOICEBOX* captures the list of items offered in the input form. Each item is marked by the element called *ITEM*. In line 146 there is an attribute called *selected*. Since this attribute is specified by the value *true*, this input field is initialized by the first business transaction pattern.

Figure 6–13 shows the result of the code described above. The tab called *General* as defined in the WDL input file displays the six worksheet entries. As we can see the first entries are write protected and their style is customized according to the attribute values.

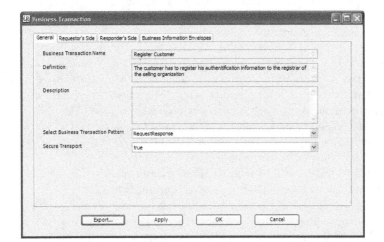

Fig. 6–13 The worksheet editor of a business transaction

As explained before, there is another XML-File for the binding of the stereotypes to the WDL instances. This file is called *DefaultWorksheets.xml* and is stored in the XML folder of the UMM Add-In. The following code represents the concept of this file. The code is logically divided into the three views *BDV, BRV* and *BTV*. The standalone-tag *STEREOTYPE* in line 173 contains the attributes *name, file* and *folder*. The attribute *name* defines the name of the stereotype, the attribute *file* gives the information about the filename and the attribute *folder* references to the *FOLDER* element inline 168 where the correct path is specified as an attribute. E.g. if the modeler wants to edit the worksheet for the package stereotyped as *business area*, the element in line 174 gives the instructions about the location of the correct WDL input file. In this case the name of the file specified by the attribute *file* is *BusinessArea.xml* and is stored in the folder called *BDV*.

The task of the file Default-Worksheets.xml

Listing 6–5 The
"deployment file" for
binding worksheets to
stereotypes

```
[165] <?xml version="1.0" encoding="utf-8" ?>
[166] <DEFAULT_WORKSHEET>
[167]
[168]     <FOLDER path="BDV" id="BDV"/>
[169]     <FOLDER path="BRV" id="BRV"/>
[170]     <FOLDER path="BTV" id="BTV"/>
[171]
[172]     <!--Stereotypes of the BDV -->
[173]     <STEREOTYPE name="BusinessDomainView"
                   file="BusinessDomainView.xml" folder="BDV"/>
[174]     <STEREOTYPE name="BusinessArea"
                   file="BusinessArea.xml" folder="BDV"/>
[175]     <STEREOTYPE name="ProcessArea"
                   file="ProcessArea.xml" folder="BDV"/>
[176]     <STEREOTYPE name="BusinessProcess"
                   file="BusinessProcess.xml" folder="BDV"/>
[177]     <!--Stereotypes of the BRV -->
[178]     <STEREOTYPE name="BusinessRequirementsView"
                   file="BusinessRequirementsView.xml" folder="BRV"/>
[179]     <STEREOTYPE name="BusinessProcessView"
                   file="BusinessProcessView.xml" folder="BRV"/>
[180]     <!--      [...]      -->
[181]     <!--Stereotypes of the BTV -->
[182]     <STEREOTYPE name="BusinessInteraction"
                   file="BusinessTransaction.xml" folder="BTV"/>
[183]     <STEREOTYPE name="BusinessChoreography"
                   file="BusinessCollaboration.xml" folder="BTV"/>
[184] </DEFAULT_WORKSHEET>
```

6.4.3   Saving the worksheet information

In order to support consistency between the UMM model and the work-sheets, worksheet information is saved as tagged values within the model. Later on, the modeler can use the export functionality for creating a documentation of the business domain knowledge. During a modeling session the modeler can use this export feature without saving the information to the tagged values. Thus there must be an internal way of handling this information already entered to the worksheet editor. Figure 6–14 shows the different ways of saving the information.

Different ways of saving the
worksheet information

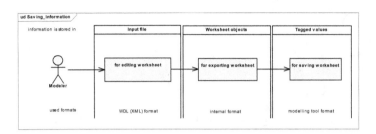

Fig. 6–14 Saving the information into different kind of formats

This figure represents the workflow of saving the information into different kind of formats. First the modeler uses the WDL file for the representation of the initial worksheet data. While the modeler enters the business domain knowledge and uses extension features of the worksheet editor (e.g. export functionality) the data should not get lost. Thus the next step visualizes the internal storage format. The last step in this workflow saves the worksheet entries as tagged values.

6.4.3.1   WDL (XML) format

When the modeler uses the worksheets in order to capture the business knowledge, he can decide between two ways of loading the WDL input file. The first one is to use the default input files for the stereotypes defined in the deployment file (see chapter 6.4.2). These files are not loaded with any initial values. But there is also a second alternative for loading the initial rendering of the worksheet editor. By specifying the correct path and filename of the WDL file, the modeler can use his own input files. These input files can contain the information about initial values of stereotypes. Since the modeler may use the export functionality to export worksheets into WDL files, he is able to save information to an external format. This external format could serve as a backup of the worksheet he was editing or as a utility to reuse this worksheet for other stereotypes. The first box in Figure 6–14 shows the scenario of saving and using the external WDL file.

Saving and reusing worksheet information with the help of the external WDL input files

6.4.3.2   Internal format

After the modeler has loaded the input file successfully, the worksheet editor displays all initial values of the worksheet entries correctly. Now he can start entering the information about the stereotype. Once he has finished, he can use several features of the worksheet editor. For example he can export all entries to an external format like Microsoft Word or he can generate UMM diagrams or elements out of the actual worksheet. While using these functions the worksheet information should not get lost. Thus

Worksheet objects represent the worksheet editor data structure

there is an internal structure representing all worksheets of the UMM model. Figure 6–15 shows the class diagram of the internal structure of the whole worksheet editor information.

Fig. 6–15 Class diagram of the internal structure of the worksheet editor

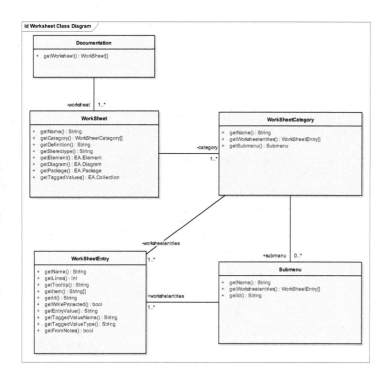

The *Documentation* class is the root-class. It keeps all the information about the worksheets. This class only takes effect, if the modeler generates a documentation of the whole model. In this case a method is parsing the whole UMM model to find every stereotype which is represented by a worksheet. These worksheets are stored as an *array* in the class variable named *worksheets*.

The Documentation class

The *WorkSheet* class is the most important class. This class represents a worksheet of exactly one stereotype. In the modeling tool Enterprise Architect, the stereotype provides methods (e.g. there is a method for returning all tagged values of a model element or displaying all links to other model elements). Thus this class implements all these methods. The function for returning all tagged values of the stereotype or the model elements representing instances of this stereotype is called *getTaggedValues()*. Further-

The WorkSheet class

more information about the type of the model element could be explored by using the method *getElement()*. The main task of this class is to store the content of the worksheets and of the design attributes for displaying the worksheet editor. As shown in the class diagram the *WorkSheet* class keeps the set of categories. This relation is represented as an association between the *WorkSheet* class and the *WorkSheetCategory* class. Furthermore there must be at least one category.

The *WorkSheetCategory* class stores the information about the different categories of a worksheet. In addition to the name of the category the *Submenu* objects and the *WorkSheetEntry* objects are stored in the instances of this class as well. Whereas the occurrence of the submenu is not mandatory, there must be at least one worksheet entry under each category. A submenu is visualized as a separated canvas in the worksheet editor representing specific input fields.

The WorkSheetCategory class and the Submenu class

The *WorkSheetEntry* class has the most attributes and methods. Every factor influencing the style of an entry is stored this class. Compared to the WDL schema file, the tags and attributes of the *ENTRY* element have the same meaning and task like the methods implemented by the *WorkSheetEntry* class. They instruct the worksheet editor, how each input field should look like. Furthermore this class keeps the information about the protection state of the input field, the number of lines, the type, the help functions and additional methods which influence the design of the input fields.

The WorkSheetEntry class

6.4.3.1   Tagged values

After the modeler has added the business information he needs to store the data captured in the worksheet editor. In order to provide an efficient and UMM compliant model there are mandatory tagged values which have to be filled out. For reasons of increased usability, the worksheet editor does this job for the modeler. Since the tagged values have to be filled out anyway, the worksheet editor uses this stack to store the business domain knowledge the modeler entered.

Saving the worksheet content into tagged values

In order to understand the power of tagged values, the usage and the advantages of this mechanism will be described. Since UML provides the *extension mechanisms* package, a coherent set of extensions for specific purposes is defined. Figure 6–16 shows this subpackage of UML that mandates how specific UML model elements are customized and extended according to specific semantics by using stereotypes, constraints, tag definitions, and tagged values. The UML provides a rich set of modeling concepts and notations that have been designed to meet the needs of typical software modeling projects. Users may sometimes require additional features beyond those defined in the UML standard. These needs are met in UML by its

The UML Extension Mechanisms provide a set of extensions for specific purposes

built-in extension mechanisms that enables the customization of modeling elements.

Tag definitions specify new kinds of properties that may be attached to model elements. The actual properties of individual model elements are specified using tagged values. They may either be simple datatype values or references to other model elements. Tag definitions can be compared to meta attribute definitions while tagged values correspond to values attached to model elements. They may be used to represent properties such as business process information (*beginsWhen, endsWhen, participants,...*).

Tag Definitions and Tagged Values

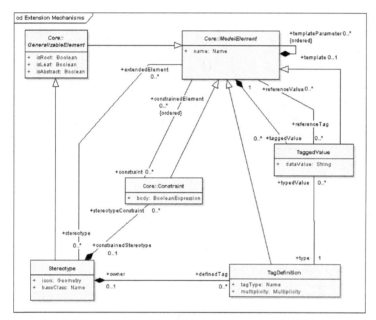

Fig. 6–16 UML - Extension Mechanisms [UMa04]

If the modeler saves the information from the worksheet editor by clicking the OK button, the UMM Add-In retrieves all the existing tagged values. If there exists a tagged value with exactly the same name as the name of an input field, the value of the tagged value will be replaced by the value of the input field. If the tagged value does not exist yet, a new tagged value will be created. Now the whole information of the worksheet is stored directly into the UMM model. After saving the UMM project in the modeling tool, the modeler can open this information the next time he wants to edit this worksheet again.

More extensions

As shown in the figure above there exists a class called *Constraint*. The worksheet editor takes advantage of this extension. In UML a constraint

Using UML constraints for saving worksheet content

may be attached to any model element to describe its semantics. A constraint which is attached to a stereotype must be observed by all model elements branded by that stereotype. Such *constraints* could be conditions like a *post-condition* or a *pre-condition*. If the WDL input file for a specific stereotype marks an input field as a constraint, the information will not be stored as a tagged value but as a UML constraint. Enterprise Architect displays both concepts in different windows representing the worksheet information. Figure 6–17 shows these two different information windows of the modeling tool.

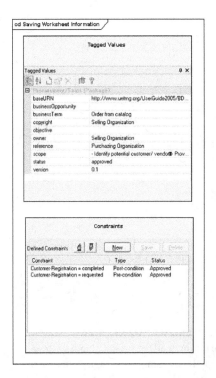

Fig. 6–17 Tagged Values vs. Constraints

Another alternative to store the information of a worksheet is to use internal storage fields of the Enterprise Architect. The modeling tool provides a lot of text areas where the modeler could store information about the model. These *note* fields are suitable for storing descriptions of worksheets. Thus if the type of an input field is marked as *notes*, the content of this input field is stored into this text area provided by the modeling tool.

Using internal storage fields provided by the modeling tool

6.5 Extensions of WDL input files

As described in the last chapters the design of the worksheets is specified by the WDL input files. But in some specific cases there are stereotypes
requiring distinctive treatment. These extensions are needed if the information is calculated rather than such exceptions are not kept in the worksheet
definition language files. E.g. displaying information about included packages of a model element is not specified by the WDL input file. This information is required in the business domain view. An extension is a hard
coded way for showing dynamical content of a model element required during runtime. In this chapter the special stereotypes are listed and these
extensions for displaying all kind of worksheets with their different characteristics will be described.

Some stereotypes require distinctive treatment

6.5.1 Business Domain View package

Displaying the business areas

The purpose of the main view *business domain view* package is classifying the business processes. The UMM meta model defines a couple of
packages beneath the business domain view. One of these packages is the
stereotype *business area*. Business areas usually follows the recommendations of the Common Business Process Catalog (CBPC). They give information about the business sections the processes are relating to [CBP03]. It
is important to have information about the occurrences of such business
areas in the worksheets. The eight normative categories of such business
areas are listed as follows.

Special characteristics of the BusinessDomainView package

- Procurement/Sales
- Design
- Manufacture
- Logistics
- Recruitment/Training
- Financial Services
- Regulation
- Health Care

If the modeler wants to edit the worksheet of the package stereotyped as
business domain view he will get information about the different types of
business areas as well. There is tab which displays a list of the names of all
included business areas. The text fields are write protected and can not be
changed within the worksheet editor. Furthermore the modeler has the possibility to add new packages for creating new business areas. Thus there is

*Displaying the list of all business areas and the pos-
sibility to generate new packages with the business matrix*

another tab for displaying a method to add new packages. This method is called "business matrix generation" and offers a couple of check boxes displayed as a matrix. The vertical check boxes represent the business areas and the horizontal ones represent the process areas. The process areas are subpackages of the business areas. A more detailed description about the business matrix, the purpose of the business areas and the task of the Common Business Process Catalog (CBPC) can be found in chapter 6.7.1. After selecting the business areas and process areas the modeler wants to add, the package structure will be created within the business domain view by clicking the *Generate* button. Thereby the modeler saves a lot of time, because he does not need to create each package manually. Furthermore the correct stereotype is added automatically to the generated packages.

Displaying the process areas

As described in the last paragraph, the business area contains other subpackages stereotyped as *process areas*. These subpackages represent the five successive phases of business collaboration defined by the ISO Openedi model [OER95]. The following classification in regard to process areas is described in the Common Business Process Catalog (CBPC).

The business area package shows the list of all included process areas

- Planning
- Identification
- Negotiation
- Actualization
- Post-Actualization

At least one of these phases has to be a subpackage of a business area. Thus there is an additional tab in the worksheet editor for showing a list of included packages stereotyped as *process area*.

6.5.2   Business Requirements View package

6.5.2.1   Displaying the subpackages of the BRV

The second package is the *business requirements view* package. This package identifies possible business collaborations and details the requirements of these collaborations [Hof05]. The UMM meta model defines five subpackages in order to capture the requirements. These packages are subpackages of the business requirements view and have the following stereotypes.

Displaying the subpackages of the business requirements view

- Business Process View

Business Entity View
Collaboration Requirements View
Transaction Requirements View
Collaboration Realization View

Every subpackage can occur several times. In order to provide a well-structured overview for the modeler, there is a tab in the worksheet editor called *Included Packages*. This tab displays the names of the subpackages grouped by their stereotypes.

6.5.2.2   Displaying extensions of the business collaboration use case

The worksheets for business collaborations in the *collaboration requirements view* display the information about relations between use cases as well as relations between use cases and their participating roles. It is important to know which use cases include other use cases and which authorized roles participate in these use cases. This information must be documented in the worksheets. Since we know that a *business collaboration use case* can include other use cases from the same stereotype or other *business transaction use cases*, there is an extra tab within the worksheet editor called *Actions* for representing this information. The tab includes an input field for a short description. Below there is a list of all included collaborations and transactions grouped by the name of the included use cases. The roles participating in these included use cases are listed within each group box. Furthermore the mapping between roles - as described in chapter 7.2.3 - is visualized using *mapsTo* relations. This scenario is represented in Figure 6–18.

Documenting the participating roles and the included use cases of a business collaboration use case

There is also one more extra tab called *Participating Roles*. All roles which are participating in the corresponding business collaboration use case are listed in this category. By clicking the *Add Role* button the UMM Add-In generates a further authorized role, which participates in this use case. The new role is directly added to the use case diagram, so the modeler saves time by being released from the burden of creating the roles manually.

Adding a role to the business collaboration use case

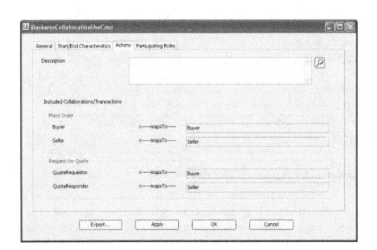

*Fig. 6–18 Included
business collaborations
and transactions*

6.5.3   Business Transaction View package

Generating a business transaction automatically

The BusinessTransaction activity graph is located in the *business interaction view* package. Since diagrams cannot be stereotyped by the modeling tool Enterprise Architect, the worksheet of a business transaction is assigned to the *business interaction* class which is the parent element of the business transaction. The speciality of this worksheet is the extra tab for generating the business transaction. This form should help the modeler to draw an activity graph out of the existing artifacts. Since the business transaction depends on actors and information envelopes already added to the UMM model, the UMM Add-In provides this information in form of drop down boxes. This method helps the modeler to create a UMM compliant business transaction in a few steps. More details about the automatic generation of diagrams are described in chapter 6.7.1.

*Generating an activity
graph out of the worksheet
editor*

6.6   Once-and-only once recording of business knowledge

An advantage of binding the worksheet editor to the modeling tool is the once-and-only once recording of business knowledge. To provide a persistent UMM model, the same information must not occur more than one time. Although a UMM model is separated into 3 different views, a logical thread is starting in the business domain view and is ending in the business transaction view. Thus the worksheet information of an efficient UMM model is required in the business domain view as well as in the business

Unique evaluation to provide a persistent data mode
for UMM worksheet entries

transaction view. The solution of this problem is linking the modeling elements with each other specifying the required information only once. This method provides a persistent data model of the UMM worksheets.

This method is required in two stereotypes. The first one is the business transaction and the second one is the business collaboration. Both activity graphs are part of the business transaction view.

The transaction and the collaboration takes advantage of this concept

6.6.0.1 Business transaction

The use cases describing a business transaction are specified in the business requirements view and stereotyped as *business transction use cases*. Once the modeler has added the worksheet information to such a use case in the *BRV*, he does not need to add the same information about the transaction in the *BTV*. If we take a look at the following code of the corresponding WDL input file, there are references to these model elements. In line 187 the attribute *taggedValueType* is set to the name of the stereotype providing the information. In this case, the stereotype *business transaction* needs to align information captured in the stereotype *requesting business activity*. In order to get the correct value of the referenced model element, the attribute *taggedValueName* must have exactly the same name as the name of the tagged value of the required information. This concept requires that the specified model element must have a link between the selected worksheet element and the other model element offering the required data. This link can be an internal link, specified by the modeling tool Enterprise Architect or a hard coded link defined in the description field of the model element. The following listing instructs the worksheet editor to display business knowledge of the corresponding model element stereotyped as *requesting business activity*.

The business transaction references the business transaction use case

```
[185] <CATEGORY name="General">
[186]         <!-------- [...]-------->
[187]         <ENTRY type="text" lines="2" protected="true"
                   taggedValueName="RequestingBusinessActivityName"
                   taggedValueType="RequestingBusinessActivity">
[188]                 <NAME>Requesting Business Activity Name</NAME>
[189]                 <TOOLTIP></TOOLTIP>
[190]         </ENTRY>
[191]         <ENTRY type="time" lines="2" protected="false"
                   taggedValueName="timeToRespond"
                   taggedValueType="RequestingBusinessActivity">
[192]                 <NAME>Time to Respond</NAME>
[193]                 <TOOLTIP>Specify the time period that this transaction
                        must be completed within.</TOOLTIP>
[194]         </ENTRY>
```

Listing 6–6 The WDL input file for the business transaction worksheet

```
[195]              <!-------- [...]-------->
[196] </CATEGORY>
```

6.6.0.1   Business collaboration

The business collaboration references the business collaboration use case. The activity graph in the business transaction view needs information of the business requirements view. In this case, the tagged values for the definition, the purpose and the description references the tagged values of the business collaboration use case. The following code lines visualize the same principle as described for the business transaction. As we can see in line 199 the attribute *taggedValueType* references the *business collaboration use case*.

The business collaboration references the business collaboration use case

```
[197] <CATEGORY name="General">
[198]              <!-------- [...]-------->
[199]              <ENTRY   type="text" lines="2" protected="true"
                            taggedValueName="definition"
                            taggedValueType="BusinessCollaborationUseCase">
[200]                  <NAME>Definition</NAME>
[201]                  <TOOLTIP></TOOLTIP>
[202]              </ENTRY>
[203]              <ENTRY   type="text" lines="2" protected="true"
                            taggedValueName="purpose"
                            taggedValueType="BusinessCollaborationUseCase">
[204]                  <NAME>Purpose</NAME>
[205]                  <TOOLTIP></TOOLTIP>
[206]              </ENTRY>
[207]              <ENTRY   type="text" lines="4" protected="true"
                            taggedValueName="notes"
                            taggedValueType="BusinessCollaborationUseCase">
[208]                  <NAME>Description</NAME>
[209]                  <TOOLTIP></TOOLTIP>
[210]              </ENTRY>
[211] <!-------- [...]-------->
[212] </CATEGORY>
```

Listing 6–7 The WDL input file for the business collaboration worksheet

6.7   Extended features of the worksheet editor

The main task of the worksheet editor is keeping the business domain knowledge in the UMM model. Before the modeler enters the documentation, he needs to add the model elements themselves. The process of drawing diagrams and adding stereotypes to the right packages requires a lot of steps. Instead of moving every single stereotype either to the drawing canvas of the modeling tool or to the tree view of the model element manage-

Supporting the modeler by generating model elements

ment, the creation of model elements should be supported by a few clicks in the worksheet editor.

Thus the worksheet editor has another purpose. In order to provide a UMM compliant model the modeler must ensure, that the structure of the model complies to the rules of the meta model. Therefore a model generation feature integrated to the worksheet editor supports the modeler in the initial phase of modeling phase. The *Matrix Package Generation* and the *Pattern Generator* provide such features. The former creates a classification structure by adding packages to the business domain view. These packages are *business* and *process areas* as defined in the Common Business Process Catalog (CBPC).

The second feature supports the modeler in creating *business transactions*. We know, that there are different kinds of patterns for business transactions. This means that the activity graph for all business transactions look the same. They only differ in the names of the activities and the information envelopes. This information is captured in the worksheets. Thus we are able to automatically generate the activity graphs from the worksheet information.

Generating diagrams by drawing different kinds of patterns

Another additional feature of the worksheet editor is the export functionality. As outlined in the beginning of this chapter, the communication between the modeler and the business domain expert is based on tables in a word processing format. Therefore the modeler must be able to export the worksheet information to such tables.

Exporting documents to word processing formats

6.7.1   Generating UMM model elements

Automatically generating UMM diagrams and structures increases the usability of the modeling process. Creating diagrams takes time. If diagrams may be computed from worksheet information, the creation of the diagrams degrades to a routine task. Inasmuch it is a wise idea to save the modeler's time by automatically generating these diagrams.

6.7.1.1   Matrix package generation

One of the first issues in the modeling process is to create a classification structure within the BDV. While collecting the artifacts, it is important to know which business process belongs to which business category. The business categories used in UMM are business areas and process areas. They are presented as stereotyped packages. The combination of business areas and process areas is called a business matrix. The packages for the business areas are direct children of the business domain view package. The packages for the process areas are subpackages of the business areas. Each busi-

Creating the initial UMM structure while defining the business areas in the BDV ness area can contain more than one process area. Business processes are assigned to process areas.

The Common Business Process Catalog (CBPC) defines a set of business and process areas with specific names. A common business process is a business process independent of any industry specific context and that may be used by a variety of companies or organizations to achieve a similar business result. The generic nature of a common business process enables one to reuse that process across different vertical industries when industry specific context and business rules are defined for a variety of e-business application integrations. The industry specific rules will be reflected in the modified attributes of the common business processes.

The Common Business Process Catalog (CBPC)

Thus the Common Business Process Catalog (CBPC) provides a framework for the analysis and reuse of common business processes [CBP03]. Furthermore it is:

- a *business library* containing all details of commonly used business processes and related information,
- a *knowledge base* for publishing and finding all business processes for registered trading partners and
- a *business service* with functions and features of UN/CEFACT core vision of doing e-Business.

The classification of business processes is inevitable in order to enable potential users to readily identify and reuse processes that might meet their business needs. UN/CEFACT recommends using the CBPC to classify business processes. The categorization schema of the CBPC is constructed within the BDV using combinations of business and process areas.

Business areas are represented as normative categories, which are reflecting the business processes at its most general level. The eight normative categories are visualized as packages in Figure 6–19.

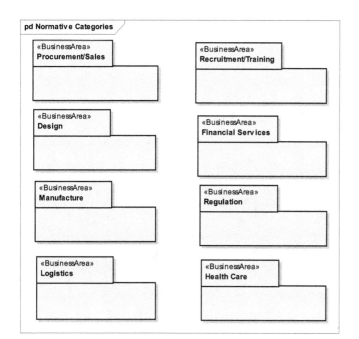

Fig. 6–19 The eight
normative categories of
the business areas

After dividing the BDV into these eight high-level business areas, it is possible to decompose each one of those areas further into process areas. There is a second-level classification in the CBPC, which defines the five successive phases of business collaboration specified by the ISO Open-edi model. Figure 6–20 illustrates these phases represented as packages stereotyped as process areas.

*Second-level classification
into five successive phases
of business collaboration*

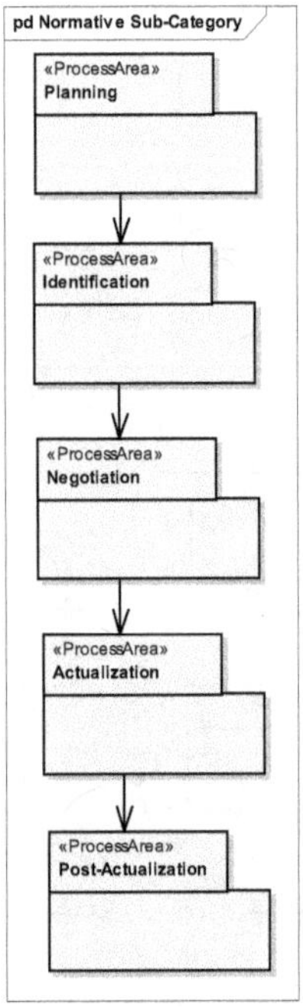

Fig. 6–20 The ISO Open-
edi phases of a business
collaboration

```
pd Normative Sub-Category

  «ProcessArea»
  Planning




  «ProcessArea»
  Identification




  «ProcessArea»
  Negotiation




  «ProcessArea»
  Actualization




  «ProcessArea»
  Post-Actualization
```

To have a better understanding of these ISO phases of a business collaboration there is a description of each package in the following paragraphs
[OPE01]:

Planning: In the *planning phase*, both the buyer and seller are engaged   *Planning phase*
in activities to decide what action to take for acquiring or selling a good,
service, and/or right.

Identification: The *identification phase* pertains to all those actions or   *Identification phase*
events whereby data is interchanged among potential buyers and sellers
in order to establish a one-to-one linkage.

Negotiation: The *negotiation phase* pertains to all those actions and   *Negotiation phase*
events involving the exchange of information following the Identifica-
tion Phase where a potential buyer and seller have (1) identified the
nature of good(s) and/or service(s) to be provided; and, (2) identified
each other at a level of certainty. The process of negotiation is directed
at achieving an explicit, mutually understood, and agreed upon goal of
business collaboration and associated terms and conditions. This may

include such things as the detailed specification of the good, service, and/or right, quantity, pricing, after sales servicing, delivery requirements, financing, use of agents and/or third parties, etc.

Actualization: The *actualization phase* pertains to all activities or events necessary for the execution of the results of the negotiation for an actual business transaction. Normally the seller produces or assembles the goods, starts providing the services, prepares and completes the delivery of good, service, and/or right, etc., to the buyer as agreed according to the terms and conditions agreed upon at the termination of the Negotiation Phase. Likewise, the buyer begins the transfer of acceptable equivalent value, usually in money, to the seller providing the good, service, and/or right.

Actualization phase

Post-Actualization: The *post-actualization phase* includes all of the activities or events and associated exchanges of information that occur between the buyer and the seller after the agreed upon good, service, and/or right is deemed to have been delivered. These can be activities pertaining to warranty coverage, service after sales, post-sales financing such as monthly payments or other financial arrangements, consumer complaint handling and redress or some general post-actualization relationships between buyer and seller.

Post-actualization phase

With this two-level categorization scheme through the eight business areas of a value chain and the five process phases of ISO Open-edi each business process can be assigned into 40 different examples of business classification. Thus large UMM models with a lot of different business areas would lead to a huge amount of packages to add. The *business matrix generation* of the worksheet editor generates these business and process areas automatically. In order to understand why this classification is comparable to a matrix, in Table 6–8 every constellation of packages is visualized in the format of a *business matrix*. Business process areas are named by concatenating the business areas names in the left column with the process area names in the top row, thus giving business process area names like *procurement/sales.negotiation* or *logistic.actualization*. Modelers may identify their initial entry points by finding the group of cells that most closely resembles the set of business processes they wish to accomplish in their proposed collaborations.

The 40 different examples of process areas

Tab. 6–8 Business Process Matrix of the Common Business Process Catalog (CBPC)[CBP03]

PROCESS AREA

BUSINESS AREA	Planning	Identification	Negotiation	Actualization	Post-Actualization
Procurement / Sales (Goods)	- Plan procurement and supply of raw materials or finished product	-Identify potential customer/ vendor - Provide catalog - Request pricing - Request availability	- Negotiate contract - Provide purchase order - Agree on delivery schedule	- Acknowledge delivery notice - Query delivery status - Receive goods - Obtain payment authorization - Make/Receive payment	- Discover delivery discrepancy - Return shipment - Negotiate/ Receive allowance or payment from vendor
Design	- Plan design service offering and use of design services	-Identify vendor who can supply product design work -Identify potential customer - Request examples of prior design work	- Submit bid - Negotiate Contract - Make counter-offers - Send design contract	- Receive designs - Inspect for conformance - Make payment - Receive payment	-Discover discrepancy in agreed upon designs - Negotiate / receive allowance or payment from vendor
Recruitment - and-Training	- Identify places in company where new staff or contract labor is needed. - Plan agency offerings - Plan student intake	-Identify personnel services provider -Identify training provider - Identify students - Request pricing - Request availability - Request/ Provide course details	- Negotiate contract for worker recruitment or training. -Agree on start dates	- Register for course - Provide training materials - Interview candidates - Make / Receive payment	- Discover discrepancy in agreed upon personnel characteristics or training components - Receive allowance or payment from vendor. - Revise training material
Logistics	-Plan provision and use of transport resources	- Identify vendors (like Fedex or UPS) - Identify potential customers - Request pricing - Request availability	- Negotiate contract for logistics service	- Receive logistics services -Provide shipping notice -Provide dangerous goods notice - Make / Receive payment	-Discover discrepancy in agreed upon service and performance - Negotiate logistics allowances with carrier - Receive allowance or payment
Manufacturing	- Plan manufacturing requirements - Research market for outsourced manufacturing	- Identify Provider of manufacturing services - Provide catalog - Request pricing - Request availability	- Negotiate contract for manufacturing - Agree on production schedule - Agree on quality standard	- Send outsourced manufacturing inputs - Receive outsourced manufacturing outputs - Make / Receive payment	Discover discrepancy in agreed upon quality levels - Negotiate allowance with vendor - Periodic evaluation
Financial Services	- Research market needs - Plan the provision of financial services (Insurance, Credit, Investments)	- Identify possible debt sources (banks, credit unions, etc.) - Identify financial institution with whom to purchase debt or equity financing - Identify insurance providers - Identify potential investors	- Negotiate terms of debt or terms of equity offering with capital provider - Agree insurance terms	- Accept transfer of debt instrument - Make transfers of stock certificate - Authorize and make payments & loans -Transfer funds - Pay Insurance Claims	- Review offerings & compliance - Receive monetary adjustments
Regulation	- Plan and	- Identify	- Agree time and	- Accept notice of	- Make adjustments

Exactly the same matrix as in Table 6–8 is represented in the matrix package generator of the worksheet editor. The screenshot in Figure 6–21 shows an example of a business process classification. The *order from quote*

The matrix in the worksheet editor represents the matrix of the CBPC

example uses two different business areas. The first one is the *procurement/sales* package and the second one is the *financial services* package. Within these packages, the second level classification are the *identification* and *negotiation* phase in the *procurement/sales* business area and the *actualization* phase in the *financial services* business area.

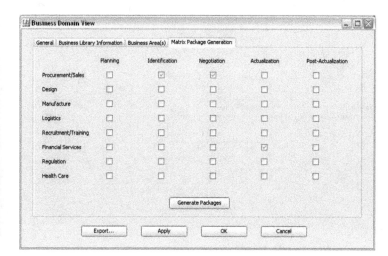

Fig. 6–21 The matrix package generation of the worksheet editor

After the modeler clicks the button for executing the matrix package generator, the packages will be added to the UMM model according to the structure as defined in the matrix. Figure 6–22 shows the generated package structure of the business domain view.

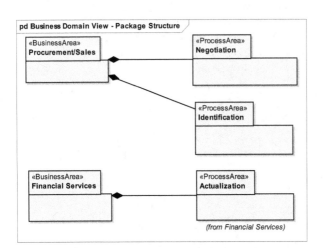

Fig. 6–22 The result of
generating the initial
package structure of the
BDV

6.7.1.1   Generating business transaction patterns

Business transactions denote an information exchange between two partners. The structure of a business transaction follows always the same pattern (see Figure 6–24 for an example). However, instances of business transactions differ from each other in regard to the exchanged information, the names of their activities and the participating roles. Therefore only a few components of the business transaction have to be altered to get a valid business transaction. It follows, that business transactions are candidate for a semi-automatic generation.

A *business transaction use case* captures the requirements and the participants of a business transaction. Hence, we decided to activate the business transaction generation wizard via the particular business transaction use case. Figure 6–23 shows the user interface of the business transaction generator.

Fig. 6–23 The user
interface of the business
transaction generator

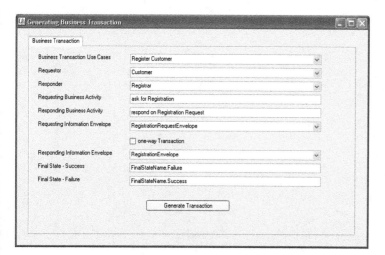

The functionality of the user
interface for generating a
business transaction

All use cases stereotyped as *business transaction use case* are listed in
this input form. The next entry is the role of the requestor. After selecting a
specific business transaction use case, the worksheet editor retrieves the
participating roles of this use case and offers it to the modeler. Since we
know that a business transaction use case can only have two participating
actors, the other actor, which was not assigned to the *requestor* is assigned
to the *responder*. The following two input fields specify the names of the
requesting and *responding business activity*. The next attribute is the
requesting information envelope. All classes stereotyped as *information
envelope* of the UMM model are listed in this drop-down box. A check box
is used to distinguish between one-way and two-way transactions. If it is not
checked, the transaction is bi-directional. In this case, the input field for the
responding information envelope is enabled and specifies the information
envelope created on the responder's side. The last two input forms define
the names for the final states. The modeler must specify a state, if the transaction succeeds and another state, if the transaction fails. After starting the
business transaction generator, the activity graph appears as a new diagram.
Figure 6–24 shows the result of generating the transaction *register customer*
described in Figure 6–23.

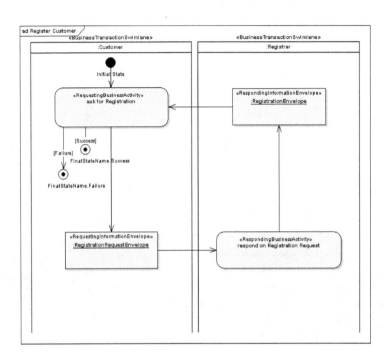

6.7.2   Exporting worksheets

The export of worksheets to different formats is an important feature of the worksheet editor. It is necessary to communicate the information kept in the worksheets to the business domain experts. Business domain experts usually do not use modeling tools. They expect plain text descriptions. Hence the business knowledge kept in the worksheets of the modeling tool must be exported to a table in a word processing format. Furthermore the export to an XML-based format stores the information outside of the model, which allows the modeler to transport the worksheet data to other models. The output formats of the worksheet editor are:

The advantages of the exporting function

- Microsoft Word – MS Office 2003
- HTML - Hypertext Markup Language
- WDL - Worksheet Definition Language (XML-based)

Different output formats

Thus the export function of the worksheet editor serves as:
- A *Text-Generator* for creating documents in a natural language
- A *Backup-System* for saving the worksheets into external files
- A *Publisher* for representing the worksheets on a Website in the Internet

A *Tool enabling Re-use* by importing the exported data into other worksheets

6.7.2.1 Export to Microsoft Word

The export function for Microsoft Word is implemented as follows: after the modeler added all the information, the input fields must be stored somewhere. Therefore the data is represented as objects of the classes shown in Figure 6–15. After these objects are filled with the worksheet information, a window appears where the modeler can specify the output format. If the radio button for Microsoft Word is enabled, the class *SelectExportFormatWindow* executes the following code lines and creates an MS Word file. In line 213 a new object for a Microsoft Word application is generated. This object is used in line 215 for instancing a new Microsoft Word document. The method *CreateTable(myWordDoc)* builds the table for displaying the worksheet information and adds it to the Microsoft Word document. Furthermore the modeler can specify the destination and the name of the output file while executing the method in line 220.

Technical implementation of the export functionality for Microsoft Word.

```
[213] Word.ApplicationClass myWordApp = new Word.ApplicationClass();
[214] object missing = System.Reflection.Missing.Value;
[215] Word.Document myWordDoc = myWordApp.Documents.Add(ref missing,
[216]                           ref missing,ref missing, ref missing);
[217] this.CreateTable(myWordDoc);
[218] myWordDoc.Activate();
[219] try {
[220]     myWordDoc.Save();
[221] } catch (Exception ex){}
```

Listing 6–9 C# code for creating a new Microsoft Word document

An important issue while implementing the Microsoft Word export functionality was the deployment of this feature. Different users have installed different versions of Microsoft Word. Thus it is difficult to communicate with the correct objects for generating the MS Word file. These objects are provided by COM (Component Object Model) technology. COM is used for the communication between applications on Microsoft platforms.

Solving the problem to communicate with the Microsoft COM (Component Object Model) technology

A solution to this problem is introduced by the .NET Framework using the concept of Primary Interop Assemblies (PIAs). In order to interoperate with existing COM types, the common language runtime requires a description of those types in a format that it can understand. The form of type information understood by the common language runtime is called metadata, and is contained within a *managed assembly*. Before an application can interoperate with COM types, it requires a metadata description of the types being used. Like any other managed assembly, an *interop assembly* is a collection

Using Primary Interop Assemblies (PIAs) for a better deployment

of types that are deployed, versioned, and configured as a single unit. Unlike other managed assemblies, an interop assembly contains type definitions of types that have already been defined in COM. These type definitions allow managed applications to bind to the COM types at compile time and provide information to the Common Language Runtime (CLR) about how the types should be marshaled at run time. While any number of interop assemblies may exist that describe a given COM type, only one interop assembly is labeled the Primary Interop Assembly (PIA). The Primary Interop Assembly (PIA) for Microsoft Office 2003 is included in the UMM Add-In installation package.

6.7.2.1 Exporting to HTML

As a second option the UMM Add-In provides an export interface for the widespread format HTML (Hypertext Markup Language). Since almost every operating system has an integrated browser, each worksheet can be viewed as a web page. The following table in Figure 6–25 shows a cutout of the exported HTML output file of a business transaction.

HTML makes each worksheet readable for everyone

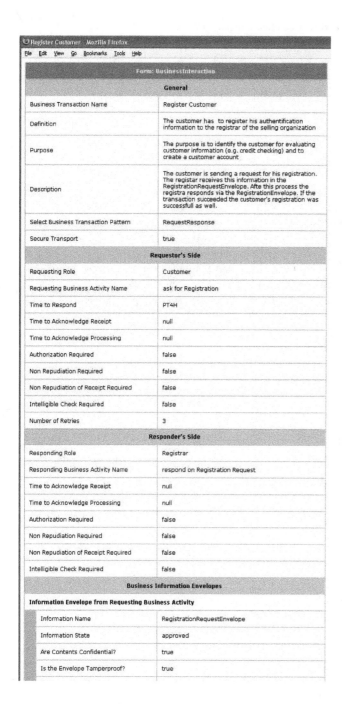

*Fig. 6–25 Example of an
HTML output of the
worksheet editor*

6.7.2.1 Export to Worksheet Definition Language (WDL)

Another opportunity to export worksheets is to use our self defined WDL formats. As mentioned in chapter 6.4.1 the worksheet editor requires an XML file for displaying the layout of the worksheet editor. Yet this file does not only define the style of the editor window, it also serves as another alternative to store worksheet information. There are specific tags in the XML structure of WDL for saving the content of the worksheet entries. At the beginning of every worksheet editing session, the UMM Add-In asks the modeler, if he wants to use the default worksheet input file, which is pre-defined for a specific stereotype or if he wants to use his own worksheet definition file. In case of using a customized input file, the modeler may use a previously exported XML file in order to reuse worksheet information.

Using the exported WDL file as an input file for the worksheet editor style

The following listing shows an excerpt of the exported WDL file for a business transaction. The business transaction worksheet has an field called *Select Business Transaction Pattern* which defines the pattern for this transaction. The tag *CHOICEBOX* in line 225 defines a drop-down box for specifying this pattern. The attribute *selected* in line 229 sets the default value of this drop-down box to the pattern the modeler specified before exporting the worksheet. Thus, when the modeler reuses this WDL input file for opening another worksheet editor window, the *request/response* pattern is selected automatically. The second entry in the code example stores the information about the responding business activity. This input field is a simple input form and has no drop down functionality. Thus the *DEFAULT* tag keeps the data of this worksheet entry.

Example of an exported WDL file

```
[222] <ENTRY type="choice" lines="1" protected="false"
           taggedValueName="BusinessTransactionPattern"
           taggedValueType="standard">
[223]      <NAME>Select Business Transaction Pattern</NAME>
[224]      <TOOLTIP>Select one of</TOOLTIP>
[225]      <CHOICEBOX>
[226]          <ITEM>CommercialTransaction</ITEM>
[227]          <ITEM>RequestConfirm</ITEM>
[228]          <ITEM>QueryResponse</ITEM>
[229]          <ITEM selected="true">RequestResponse</ITEM>
[230]          <ITEM>Notification</ITEM>
[231]          <ITEM>InformationDistribution</ITEM>
[232]      </CHOICEBOX>
[233]</ENTRY>
[234] <ENTRY type="text" lines="2" protected="true"
               taggedValueName="RespondingBusinessActivityName"
               taggedValueType="RespondingBusinessActivity">
[235]      <NAME>Responding Business Activity Name</NAME>
[236]      <DEFAULT>respond on Registration Request</DEFAULT>
```

Listing 6–10 The result of exporting the worksheet using WDL

```
[237]      <TOOLTIP />
[238] </ENTRY>
```

The output of the exported file in Listing 6–10 has the same structure as the WDL input file for a business transaction. The only difference is that the exported WDL file is initialized by the data of the worksheet editor. Thus the WDL export functionality serves as a backup system for saving the worksheet information to an external file.

7   User Guide

Worksheets in the context of the modeling process

Constructing B2B information systems requires collecting business as well as technical information concerning the target business domain. This information is the base for defining the scope of the system, its requirements as well as its specifications in further steps of our methodology. In order to facilitate this knowledge gathering process, UMM uses the concept of worksheets. Worksheets help to guide the interview process with business experts and computer engineers and grasp their know-how. Hence, in our methodology worksheets are the bridge between business experts and business analysts.

Each view and each subview requires a tailored set of worksheets in order to collect relevant information. It is noteworthy that UN/CEFACT does not standardize worksheets, but advises which information to collect. We detail these issues in own subsections pertaining to the particular (sub)view sections.

The expertise captured using worksheets is the starting point for describing collaborations and their requirements. In other words, worksheet input is transferred into model elements or assigned to tagged values. However, some information is only captured for documentation purpose and to effectuate a common standard of knowledge. This ensures that all participants have a common view on the same business domain.

As outlined above, we detail the information that should be grasped in each stage of the methodology as part of the particular (sub)view sections. General information that is applicable to a multiplicity of worksheets is only described once below and then referenced in the appropriate worksheet section.

Common Information

Name: The name of a particular element
Description: Free text information describing an element in more detail

Business library information

BaseURN: The namespace of a *business library* that is subject to a registry. The *base URN* is both the unique identifier for the *business library* itself and the namespace for the elements contained in the *business library*. The unique identifier of elements contained in a *business library* results from the *base URN* of the container and the local name of the element. The *base URN* should be assigned by the modeler under semantical considerations. The *base URN* should not be changed after the library has been registered.

Version: Holds the current version of a registered *business library*. Version information should not be assigned by the user, but it should be managed by a registry. It follows, that an unregistered library should not have version information assigned.

Status: Holds the current status of a registered *business library*. Status information should not be assigned by the user, but it should be managed by a registry. It follows, that an unregistered library should not have status information assigned.

Business term: A business term is a synonym, by which a business entity is commonly known.

Owner: The owner of the *business library*. This might be an organization, an institution or an individual.

Copyright: Holds copyright information about the *business library*.

Reference: Holds location information about continuative documentation about a particular *business library*.

7.1 Business Domain View

7.1.0.1 Overview and purpose

The *business domain view* (BDV) is the starting point for modeling business collaborations. At this early stage of our modeling methodology the characteristics, requirements and procedures of a target business domain are determined. Thus, the BDV is utilized to gather existing knowledge from stakeholders and business experts regarding business processes they participate in or just have an interest in. Discovery and identification of inter- and intraorganizational business processes on a high level is the purpose of this stage. No new collaborative business processes are constructed. The goal is to identify possible areas for business interactions between two or more business partners. A business collaboration constructed in later steps of our modeling methodology must respect the characteristics of the

The business domain view discovers, but does not construct collaborative business processes

business processes identified in the BDV and must not be in conflict with them.

Although UMM focuses on defining and describing B2B processes, intraorganizational processes are also documented in the BDV work steps. Usually domain experts describe business processes from a partner specific view. The interface of such a business process could include communication with a partner and consequently the intraorganizational process would be a candidate for an entry point of an interorganizational collaboration.

BDV captures intra- and interorganizational processes

The requirements gathering process to capture the domain knowledge is primarily accomplished through interviews with business experts and stakeholders. Predefined and standardized worksheets guide the business process analyst through the interview and help capturing the knowledge of the business participants. Despite using these worksheets in the interview process for guidance, the interviewer has to make sure not to influence the respondent. It is important that the dialog between the stakeholder and the interviewer is in the language of the stakeholder, technical and modeling terms should be avoided [Hof05]. The interrogation of the stakeholders results in an overview of the business processes and the participants of a specific business domain.

Knowledge is captured through interviews with business experts

The identified processes should then be classified according to predefined and adopted classification schemes in order to enable easy and semantically correct reuse. This step is intended to reuse existing knowledge in future projects in order to save time and money. Therefore it is advisable to apply the classification process also for artifacts that are constructed later in the *business requirements view* (BRV) and in the *business transaction view* (BTV). It is recommended to utilize UN/CEFACT's *Catalog of Common Business Processes* (CBPC) as a classification scheme. Other candidates for the classification scheme are - but are not limited to - the *Supply Chain Operations Reference Model* (SCOR) or *Porter's Value Chain* (PVC).

7.1.0.2 Stereotypes

BusinessProcess (UseCase): According to Hammer and Champy a business process is defined as a flow of related activities that together create a customer value [HC93]. Business processes might be either performed by one partner (intraorganizational business) or by two or more partners (interorganizational).

BusinessCategory (Package): Business categories are used to categorize *business processes*. Either one or more *business categories* or a combination of *business* and *process areas* can be utilized for classification. A *business category*, as well as their specializations *business area* and *process area*, might be structured recursively.

BusinessArea (Package): A *business area* is a specialization of a *business category* and roughly corresponds to a division of an enterprise.

ProcessArea (Package): A *process area* is a specialization of a *business category* and corresponds to a set of common operations within a *business area*.

Stakeholder (Actor): A *stakeholder* is interested in a *business process*, but does not take an active part in its execution.

BusinessPartner (Actor): A *business partner* plays a role in the execution of a *business process*. Thereby he has a natural interest in this process.

7.1.0.3   Worksheets

Business domain view worksheet

Common Information

Business library information

Included business areas: A listing of *business areas* that are part of the *business domain view*. A *business area* is a categorization mechanism. Hence we only capture *business areas* that are needful to classify identified *business processes*. A *business area* is further described by a *business area* worksheet.

Business area worksheet

Common Information

Business library information

Objective: The purpose and value to be achieved by the *business process* within the *business area* under consideration

Scope: Defines the boundaries of the *business area*

Business opportunity: The strategic interest in the particular *business area*

Included process areas: A listing of *process areas* that are contained in a *business area*. Each *process area* in this list is further detailed by its own *process area* worksheet.

Process area worksheet

Common Information

Business library information

Objective: The purpose and value to be achieved by the *business process* within the *process area* under consideration

Scope: Defines the boundaries of the *process area*

Business opportunity: The strategic interest in the particular *process area*

Included business processes: Lists identified *business processes* that take place in a *process area*. A *business process* is further described in detail by a *business process* worksheet.

Business process worksheet

Common Information

Business library information

Definition: Describes the customer value that is created by the *business process*. If multiple parties collaborate in the execution of a *business process* the overall value is identified.

Participants: Identifies all parties that play an active role in the execution of a *business process*. Each identified participant is modeled as a *business partner* in a further step and connected to the process via a *participates* association.

Stakeholder: Holds information about the stakeholders of a *business* process. A *stakeholder* has interest in a process but plays no active role in its execution. Each identified party is modeled as a *stakeholder* and connected to the process via a *is of interest to* association.

Reference: Holds location information of further information that is relevant to the process.

Pre-conditions: Describes requirements that have to be fulfilled in order to execute a *business process* (e.g. a customer has to be registered to issue an order).

Post-conditions: Identifies conditions that have to be satisfied just after the execution of a *business process* (e.g. an order is confirmed).

Begins when: Specifies semantic conditions (business events) for the initiation of the *business process*. It may be used to specify a semantic state, that has to be reached in order to start the process (e.g. the seller's ordering process is started when the order is received).

Ends when: Specifies business events that indicate the finalization of a *business process* (e.g. the seller's ordering process is finished when the confirmation is sent).

Actions: Describes one or more activities that are performed in the execution of a *business process* (e.g. the seller checks the buyer's account and the stock of the ordered goods prior he sends a confirmation).

Exceptions: Identifies errors that may occur during the execution of a *business process* (e.g. the seller has the ordered goods not in stock)

Included business processes: A *business process* may require the execution of other *business processes* as a part of its own workflow. Each included *business process* results in an additional modeled *business process* connected via an *include* association.

Affected business entities: Identifies *business entities* that are affected by the execution of the *business process*. A *business entity* is a real-world thing having business significance (e.g. an order). *Business entities* identified in this step are input to the workflow in the *business entity view* (See "Business Entity View" on page 112).

7.1.0.4   Step by step modeling guide

The BDV involves two main steps:

1. Define a classification schema
2. Identify relevant *business processes*, their *participants* and *stakeholders*

Define a classification schema

The goal of the *business domain view* is to gather knowledge about an existing business domain. In a first step it is necessary to subdivide the business domain into groups of related business processes. In other words, we first need a categorization schema which is later used to classify the business processes. UMM offers the concept of *business areas* and *process areas* to classify *business processes*.

Define an appropriate classification for the identified business processes

Tab. 7–1 Order from quote example: worksheet for the business domain view

Form: BusinessDomainView	
General	
Business Domain View	Order From Quote
Description	This business domain describes a horizontal business process scenario where a purchasing organisation purchases from a selling organisation. It this scenario, the purchasing organisation may or may not have an existing account with the selling organisation and therefore an the establishment of this account may be required. This scenario assumes that the purchasing organisation does not have contract prices for any of the selling organisation goods or services and therefore the purchasing organisation must obtain a quote from the selling organisation prior to placing an order.
Business Library Information	
Base URN	http://www.untmg.org/UserGuide2005/BDV/OrderFromQuote
Version	0.1
Status	approved
Business Term	Purchase Order, Request for Quote, RFQ, Order, Sales Order
Owner	UN/CEFACT
Copyright	UN/CEFACT
Reference(s)	- Purchasing Organization - Selling Organization

A *business area* usually corresponds to a division of an enterprise. In order to develop a categorization schema it is necessary to identify all *business areas* of a business domain. This modeling step is supported by the *business area* worksheet (Table 7–2), which must be completed for each identified *business area*. It is also possible to identify *business areas* that are nested in another *business area*. In other words, *business areas* may be nested recursively and build a tree structure. A *business area* on the lowest level of such a tree is called a leaf *business area*.

Add business areas to the business domain view

The UMM does not mandate a predefined classification of *business areas*. However, UN/CEFACT's *Common Business Process Catalog* (CBPC) recommends a list of eight categories: *procurement/sales, design, manufacture, logistics, recruitment/training, financial services, regulation* and *health care*. This list of *business areas* is considered as non exhaustive. The recommendation in the CBPC does not use any recursive nesting, i.e. a flat list of *business areas*.

predefined business areas of the CBPC

The next step is to identify the relevant *process areas* within each leaf *business area*, where a *process area* corresponds to a set of common operations within that *business area*. *Process areas* might also be modeled recursively to represent an appropriate categorization for *business processes*. A leaf *process area* is on the lowest level of the resulting tree structure. A leaf *process area* is a category of one or more common *business processes*, which will be identified in the second main step of the BDV. Thus, a *process area* contains either other *process areas* or *business processes*.

*Fig. 7–1 UN/CEFACT's
Common Business
Process catalog matrix*

	Planning	Identification	Negotiation	Actualization	Post-Actualization
Procurement/Sales	☐	☑	☑	☑	☐
Design	☐	☐	☐	☐	☐
Manufacture	☐	☐	☐	☐	☐
Logistics	☑	☑	☑	☐	☐
Recruitment/Training	☐	☐	☐	☐	☐
Financial Services	☐	☐	☑	☑	☐
Regulation	☐	☐	☐	☐	☐
Health Care	☐	☐	☐	☐	☐

Again, UN/CEFACT does not dictate a predefined classification for *process areas*. Nevertheless, it is recommended to use the list of *process areas* identified in the CBPC: *planning, identification, negotiation, actualization, post-actualization*. This list corresponds to the five successive phases defined in the Open-edi reference model [OER95].

In a first example (Figure 7–2) we use the CBPC classification in order to categorize the business domain of our *order from quote* example. Figure 7–1 shows the full CBPC matrix.

Not all combinations of *business* and *process areas* are relevant in our example. Our example spans over three *business areas*: *financial services, logistics* and *procurement/sales*. Each of these *business areas* is described by its own *business area* worksheet. Table 7–2 shows the *business area* worksheet for *procurement/sales* of our example.

Tab. 7–2 Order from quote example: worksheet for the procurement/sales business area

Form: BusinessArea	
General	
Business Area	Procurement/Sales
Description	In this business area, business processes are described where a purchasing organisation can find potential suppliers for required products, can establish an account with the selling organisation, request for a quotation of required products and eventually place a purchase order with the selling organisation if the quote provided by the selling organisation meets the purchasing organisation's business objectives.
Objective	The objective of this business area allows a purchasing organisation to find an appropriate supplier (selling organisation), to establish an account, to request a quote for required products, and finally to purchase these products.
Scope	- Identify potential customer/ vendor - Request quote for price and availability - Request purchase order
Business Opportunity	The business opportunity of this business area is to allow purchasing organisations to purchase required products from selling organisations.
Business Library Information	
Base URN	http://www.untmg.org/UserGuide2005/BDV/Procurement
Version	0.1
Status	approved
Business Term	Purchase Order, Order, RFQ, Quote, Quotation, Sales Order, Price Request
Owner	UN/CEFACT
Copyright	UN/CEFACT
Reference(s)	·

Within *procurement/sales* the *process areas* of *identification, negotiation* and *actualization* are relevant. Each *process area* is again detailed using its own *process area* worksheet. Table 7–3 shows the worksheet for the *identification process area* of the *procurement/sales business area*. Regarding the other *business areas* of our example, *negotiation* and *actualization* are pertinent to *financial services*. For *logistics* the *planning, identification* and *negotiation process areas* are considered as relevant. The resulting combinations are marked by ✓ in the matrix of Figure 7–1. In UML *packages* must be ordered hierarchically. Thus the matrix must be represented as a tree structure as depicted in Figure 7–2. The three *business area* packages *logistics*, *procurement/sales* and *financial services* are beneath the *business domain view* package. Each of these packages includes the corresponding *process areas*.

*Tab. 7–3 Order from quote
example: the worksheet
for the process area
identification*

Form: ProcessArea	
General	
Process Area	Identification
Description	The Identification Phase pertains to all those actions or events whereby data is interchanged among potential buyers and sellers in order to establish a one-to-one linkage
Objective	
Scope	- Identify potential customer/ vendor - Provide catalog - Request pricing - Request availability
Business Opportunity	
Business Library Information	
Base URN	http://www.untmg.org/UserGuide2005/BDV/Procurement/Identification
Version	0.1
Status	approved
Business Term	Purchase Order, Order, RFQ, Quote, Quotation, Sales Order, Price Request
Owner	UN/CEFACT
Copyright	UN/CEFACT
Reference(s)	

*Fig. 7–2 Classification of
business processes
according to the CBPC*

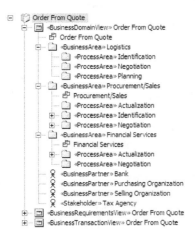

In order to demonstrate the recursive nesting of *business areas* (and   *Nesting business areas*
process areas) we use an additional example (Figure 7–3). Consider a business domain that covers the *business areas telephone services* and *internet
services*. The *telephone service* is further restructured into *fixed network,
mobile services* and *voice over IP* - each of which are nested *business areas*.
Similarly the internet services *business area* is composed of the nested *business areas private* and *business solutions*. Only these second level *business*

areas - which are the leaf *business areas* in our example - contain *process areas*.

Fig. 7–3 *Defining nested business areas*

If such a precise classification using *business areas* and *process areas* is not essential, you may use the more general classification type *business category*. *Business categories* may also be nested to obtain a hierarchical classification. In a UMM model either *business categories* or combinations of *business areas* and *process areas* are allowed to classify *business processes*.

Figure 7–4 shows a classification scheme built of only *business categories*. The example model deals with processes occurring within the area of operations of *waste management*. The *business category* on the first level is named *cross border waste shipment*. Beneath we defined two other *business categories* named *waste management notification* and *waste management transport*. All identified *business processes* would go in one of these two leaf *business categories*.

Business categories may be used instead of business area/process area combinations.

Fig. 7–4 *Composing a classification scheme of nested business categories*

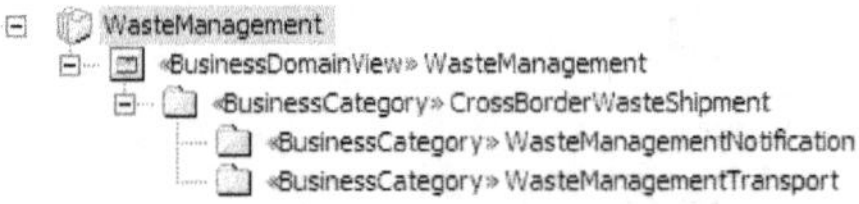

If no classification schema is known or the modeling project is rather narrow in scope, it may be sufficient to have only one *business category* package beneath the *business domain view*. All *business processes* identified in the next step will then go into this single *package*. It should be noted that UMM is an iterative process. This means the package structure is not fixed all the time. It might be changed due to new insights during the identification of *business processes* in the next step.

Identify relevant business processes, their participants and stakeholders

In this stage a partner's *business processes* are identified on a high-level and modeled in the appropriate *process area* or *business category*. A *business process* may include the execution of one or more other processes. Such a relationship is collected in the *business process* worksheet and modeled using an UML *include* relation.

Model the identified business processes

The *business process* worksheet captures the related parties of a *business process*. A related party might be either a *business partner* or a *stakeholder*. A *business partners* takes up a role in the execution of a *business process* and is denoted as a *participant* in the worksheet. The relationship between a *business partner* and a *business process* is modeled via a *participates* association. A *stakeholder* has interest in a *business process*, but does not take an active part in its execution. In order to indicate that a *business process* is of interest to a *stakeholder*, connect them via an *is of interest to* association. If *business partners* and *stakeholders* are related to *business processes* in more than one *process area*, model them in the corresponding parent *business* or *process area*.

Describe the relationship between business processes and related parties

You utilize *use case diagrams* in order to facilitate the modeling of *business processes* and their relationship with other processes and related parties. You may use only one *use case diagram* per *process area* (or *business category*) showing all its *business processes*. However, you may also use multiple *use case diagrams* per *process area* (or *business category*), e.g. showing only the *business processes* of a particular partner in each *use case diagram*.

Regarding our *order from quote* example, the following *business processes* and *business partners* are identified in the *procurement/sales business area* during the *negotiation* phase: Figure 7–5 shows the *business processes* of the *selling organization*. The *selling organization* itself is denoted as *business partner* and connected with each *business process* (*register customer* and *request credit check*) it takes part in via a *participates* association.

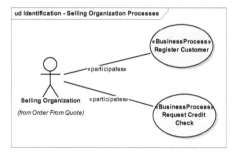

Fig. 7–5 Use cases of the selling organization's processes

Similarly, the following diagram (Figure 7–6) shows the *purchasing organization* as well as its *business process* within the *process area* identification (*get customerID*). The *purchasing organization* is again modeled as a *business partner* and connected with the process via a *participates* association.

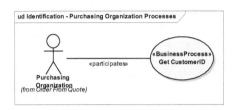

Fig. 7–6 The purchasing organization participates only in a business process named get customerID

7.1.0.1   Artifacts

Results of the *business domain view* are use case descriptions of the target business domain consisting of its occurring **business processes** and related **business partners** and **stakeholders**. *Business processes* and their *worksheets* should be described as detailed as possible by the knowledge gathered through the interview process.

7.2   Business Requirements View

7.2.0.1   Overview and purpose

The goal of the *business requirements view* (BRV) is to identify and describe the requirements of business processes involving two or more participants. Business processes with two or more participating roles are called business collaborations in the UMM. A business collaboration is further characterized by the fact that it is built up by several business processes.

In order to specify possible business collaborations, the first step is to consider the identified *business processes* from the BDV and to look for complementary processes of other partners. *Business processes* of different partners having complementary interfaces are strong candidates for being part of a business collaboration.

Identify business processes that are candidates to be part of a business collaboration

Business processes that are considered as relevant in a possible collaboration should be refined in the *business process view*. In this step a process of interest is decomposed into a flow of activities. Executing a *business process* affects *business entities* and thus changes their state. The *business process view* describes the flow of activities in a *business process* and the resulting business state changes. *Business entities* that are manipulated in relevant *business processes* as well as their behavior are modeled in *business entity views*. Since the *business process view* uses states of *business entities* defined in a *business entity view*, modeling of both views is tightly coupled.

Relevant business processes and business entities are described in detail

The *business requirements view* covers the identification of the roles involved in a certain collaboration. Business collaborations are later specified in the *business transaction view* by means of *business collaboration protocols*. In the business requirements view, requirements of a collaboration are specified in a subview called *collaboration requirements view*.

Identifying the roles participating in a process is one of the major BRV tasks

A business collaboration spans over several atomic interactions, called *business transactions* in the UMM, and other collaborations. A *business transaction* is an interaction on the lowest level of granularity and thus always performed by two roles. Each transaction role as well as each nested collaboration role needs to be mapped to exactly one role of the including collaboration. Requirements of a transaction are described in a *transaction requirements view*.

A collaboration may be performed by different sets of *business partners*. Thus there may be multiple *business collaboration realizations* of a business collaboration. In order to obtain concrete business collaboration descriptions, *business partners* taking part in a collaboration are mapped to collaboration roles in the *collaboration realization view*.

Different sets of business partners might execute the same collaboration

7.2.0.2   Stereotypes

BusinessProcessView (Package): In a *business process view* a *business process* that is relevant for a collaboration is decomposed into a flow of activities. The decomposed process may either be an internal process or a process that connects internal processes of *business partners*. The workflow of a *business process* might change the states of *business entities*. Such state changes are also captured in this view.

BusinessEntityView (Package): The *business entity view* describes *business entities* that are affected by a process including their lifecycle and state changes.

CollaborationRequirementsView (Package): The *collaboration requirements view* contains all elements that describe the requirements of a business collaboration as well as its participating roles.

TransactionRequirementsView (Package): The *transaction requirements view* contains all elements that describe the requirements of a *business transaction* as well as its participating roles.

CollaborationRealizationView (Package): The *collaboration realization view* contains all elements specifying the concrete realization of an abstract business collaboration. In this stage *business partners* are mapped to abstract collaboration roles.

7.2.0.3 Worksheets

The worksheets of the *business requirements view* are described in the continuative sections of the particular subviews of which they are part of. However, we use one worksheet to gather common information about the *business requirements view* package.

Business requirements view worksheet

- *Common Information*
- *Business library information*

7.2.0.4 Step by step modeling guide

1. Model the existing and/or desired process workflow(s)
2. Identify relevant *business entities*
3. Describe requirements on *business collaboration protocols*
4. Describe requirements on *business transactions*
5. Define concrete realizations of business collaborations

Model the existing and/or desired process workflow(s)

Modeling the *business requirements view* is straightforward, because it just acts as a container for its subviews. The first step refines *business processes* from the BDV that are of relevance for a desired collaboration. You may either refine all relevant *business processes* in one *business process view* or you may create one *business process view* per each process. Although modeling the *business process view* is strongly recommended, you may omit this modeling step. Thus zero to many *business process views* are allowed in the *business requirements view*.

Decompose business processes in the business process view

Identify relevant business entities

After the decomposition of pertinent *business processes, business entities* that may be of relevance to a collaboration are further described. One *business entity* view may be the container for all identified *business entities* or each *business entity* may be modeled in its own *business entity view*. UN/CEFACT recommends, but does not mandate, to gather *business entities* having business significance.

Business entities and their life cycle are described in business entity views

Describe requirements on business collaboration protocols

The next step in the BRV workflow is describing requirements and participating roles of a business collaboration. Each business collaboration must be described in its own *collaboration requirements view.*

Collaboration requirements are described in collaboration requirements views

Describe requirements on business transactions

Similar as for business collaborations, each *business transaction* and its participating roles must be specified in their own *transaction requirements view.* Hence a UMM model exists of one *collaboration requirements view* per collaboration and one *transaction requirements view* per transaction.

Requirements of transactions are specified in transaction requirements views

Define concrete realizations of business collaborations

Finally, each collaboration needs to be realized by at least one concrete collaboration realization. Each realization has to be specified in its own *collaboration realization view.* No other model elements except the five subviews described above are allowed in the *business requirements view.*

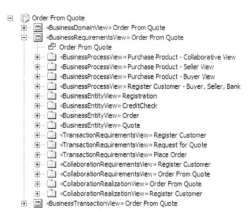

Fig. 7–7 BRV package structure of the order from quote example

Figure 7–7 shows the *package* structure in the *business requirements view* concerning the *order from quote* example. Process flows are described in four *business process views.* Each identified *business entity* is kept in its own *business entity view* (but modeling all *business entities* in one *business entity view* is also possible). Furthermore, there exist two *collaboration requirements views* - one for the *register customer* and one for the *order from quote* collaboration - that specify the requirements on the particular collaboration. The same applies for the *transaction requirements views*, which captures the specifications of *business transactions.* Finally, we have two *collaboration realization views* in our example - one for the *order from quote* and one for the *register customer* collaboration.

A business collaboration may have an unlimited number of concrete realizations. Each different set of *business partners* that performs the same collaboration entails another concrete realization of an abstract collaboration. In other words, each occurrence of the same collaboration with a different set of participants must be manifested in its own *collaboration realization view*.

7.2.0.1 Artifacts

The *business requirements view* facilitates identifying collaborations and collecting requirements on them. In order to get an overlook of the workflow of business processes that are relevant for a collaboration, these business processes are decomposed into a flow of activities. The internal flow of activities of a *business process* is modeled in **business process activity models**.

Furthermore, relevant *business entities* are identified and their life cycles are specified using **business process entity life cycles**.

Business collaboration use cases describe identified collaborations and their relationships to included **business transaction use cases** and other *business collaboration use cases*. A *use case* is described by its corresponding worksheet, associations to its participating roles and roles mappings to included *use cases*. **Business collaboration realizations** specify realizations between a specific set of *business partners* of the rather abstract *business collaboration use cases*. *Business partners* participating in a *business collaboration realization* must be mapped to the *authorized roles* of a *business collaboration use case*.

7.2.1 Business Process View

7.2.1.1 Overview and purpose

The *business process view* covers business processes that may be candidate for business collaborations. It is the purpose of this step to get a deep understanding of relevant business processes.

Business processes may either be adopted from the BDV or newly constructed. Constructing processes is required if a desired collaboration needs additional work units which were not described in the BDV. In order to analyze the activities and the participants of a process, a *business process* is decomposed into a *business process activity model*. If a *business process* is decomposed or not depends on its relevance and complexity.

A business process is decomposed into a flow of activities

A *business process activity model* may either describe an internal process of a partner or a collaborative process between partners. In the case of
an internal process, the process flow is considered to discover interfaces
requiring interaction with a business partner. These interfaces of the internal
processes have to be minded in future collaborative processes.

Business processes are examined to discover required interactions between partners

Modeling collaborative processes helps understanding the workflow of
a process and the interaction between business partners in this process.
Besides the activity flow, *business process activity models* describe how the
workflow of a process affects the states of *business entities*. *Business entity
states* are the output of *business process activities* and input to following
activities. The state of a *business entity* is dependent on its preceding activities and influences in turn the further workflow.

We distinguish *internal* and *shared business entity states*. *Internal business entity states* occur just in the internal workflow of a business partner.
Shared business entity states denote interfaces between two partners that
require a state change in the systems of both partners. It follows, that *shared
business entity states* should be considered as candidates for interactions
between partners in further modeling stages.

Using *business process activity models* in the BRV facilitates the construction of collaborations and transactions in a later design step.

7.2.1.2   Stereotypes

Business Process (UseCase): A *business process* is a flow of related
activities that together create a customer value. Business processes
might be either performed by one partner (interorganizational business)
or by two or more partners (intraorganizational).

BusinessProcessActivityModel (ActivityGraph): A *business process
activity model* represents a flow of a *business process*. They are means
to describe and understand (collaborative) processes and help to discover interfaces for connecting processes of different partners.

BusinessProcessActivity (State): A *business process activity* is a step in
the workflow of a *business process*. A *business process activity* might
be refined by another *business process activity model*.

InternalBusinessEntityState (ObjectFlowState): An *internal business
entity state* is a state of a *business entity* that is just of relevance to a single business partner.

SharedBusinessEntityState (ObjectFlowState): A *shared business entity
state* is a state of a *business entity* that is of relevance to two ore more
business partners. *Shared business entity states* require interactions

between the involved partners, to synchronize the states of the affected *business entities*. Hence, they indicate the need for an interaction between those partners.

7.2.1.3   Worksheets

Business process view worksheet

Common Information
Business library information

7.2.1.4   Step by step modeling guide

1. Decide on the structure of the *business process* descriptions
2. Model partitions when describing an interorganizational process
3. Describe the flow of activities
4. Identify *business entities* and their *shared* and *internal states*

The *business process view* does not have strict modeling guidelines. There exist several ways for describing business processes of interest and analyzing their flow of activities. Describing the process flow is not supported by worksheets. The business analyst rather works along with the business expert in the *business process view*, because it is easier to describe a process by graph than by words.

Decide on the structure of the business process descriptions

As first step you have to decide how to structure the *business process view*. You may either model each *business process* in its own *business process view* or you may refine several *business processes* together in one *business process view*. Processes that are described in one view, should in some way relate to each other. Anyway, information about the *business process view package* and its contents is gathered through the *business process view* worksheet. The *business process view* worksheet corresponds to the worksheet of any other *business library*. An example worksheet for a *business library* is given by the worksheet of the *business domain view package* (Table 7–1).

Relevant business processes are subsequently described in detail. If a *business process* is constructed from scratch in the BRV, add a *business process* to the model and optionally refine it with a *business process activity model*. If a process identified in the BDV is described, you may choose one of the following two alternatives: you may either drag the *business process* from the BDV to your *business process view* (and illustrate it with a *business process activity model*) or add only the refining *business process activ-*

The business process view may be structured in different ways

Model the business process that is going to be refined

ity model to the *business process view*. Independent of the two alternatives, UN/CEFACT encourages using *business process activity models* to gain a deep understanding of pertinent business processes.

Model partitions when describing an interorganizational process

In the next step a *business process* is decomposed in a *business process activity model*. If the refined process is a collaborative process, UML *partitions* should be used to collate *business processes* to *business partners*. In order to show that activities in a certain *partition* are performed by a *business partner*, the *business partner* has to be assigned as a *classifier* to the *partition*. For each *business partner* participating in the process, a *partition* must be added to the *business process activity model*. If an internal process is refined, you may add a *partition* or not. Anyway, if a *partition* is added for an internal process, the corresponding *business partner* must be added as the *classifier* of the *partition*.

Use partitions to assign activities of a business process to a certain partner

Describe the flow of activities

Regardless if *partitions* are used or not, the sequence of the business processes has to be decomposed into *business process activities* and *business entity states*. For each step in a business process a *business process activity* has to be added. If a *business process activity* is decomposed into further activities, refine it by another *business process activity model*.

Denote a workflow step by a business process activity

Figure 7–8 shows an example *business process activity model* describing the *purchase* process (from the *order from quote* example) seen from the buyer's internal perspective. Hence, an intraorganizational process is illustrated and no partitions are required. Furthermore, the *business process activity model* contains only a flow of *business process activities* describing the buyer's *purchase* process in detail.

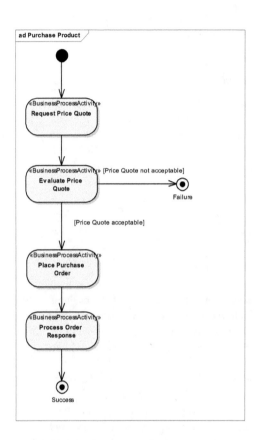

Fig. 7–8 Purchase
process viewed from the
buyer's internal
perspective

Identify business entities and their shared and internal states

A *business process activity* may affect the state of a *business entity.* Indicate this by modeling a business entity state as the successor of a *business process activity.* If a business entity state is just meaningful for one partner, add an *internal business entity state* to the *partition* (or directly to the *business process activity model* if no *partitions* are used). Business entity states that require an interaction between two partners are denoted as *shared business entity states* and placed between the *partitions* of the interacting partners.

Describe changes of a business entity state

Internal and *shared business entity states* are instances of *business entities* in a certain object state. At this stage, *business entities* are not modeled yet, hence it is advised to switch between *business process view* and *business entity view* during the modeling process. Start with the *business process view* and describe *internal* and *shared business entity states* initially

Switching between business process view and business entity view may be required in the modeling workflow

without a *classifier*. Then you should continue with the *business entity view* (see next section) and describe *business entities* together with their life cycles. Finally, switch back to the *business process view* and set the *classifier* and *state* of each object using *business entities* and their lifecycle descriptions.

However, in the *business process view* it is not mandatory to identify and model *business entities* and their state changes. As Figure 7–8 depicts, *business process activity models* can be used in an early modeling stage to get an understanding of a business process workflow. In such early elaboration steps, business entities may have not been identified yet. Nevertheless designing a desired collaborative process workflow should involve the identification of *business entities* and how they interact with atomic *business process activities*.

Identifying business entity states is recommended, but not mandatory

Taking a look at our *order from quote* example, Figure 7–9 shows an extract of the *business process activity model* describing the *register customer* process. Compared to the internal view on the *purchase* process (Figure 7–8), this diagram shows an inter-organization process workflow. Hence, we use two *partitions* in this extract - one for the *purchasing* and one for the *selling organization*. Furthermore, *classifiers* of *partitions* refer to *business partners* who execute the business activities in the particular *partition*. In the example, the *purchasing organization* performs the *get customer ID* activity and the selling organization executes *act on registration request* and *request credit check*. Between the two *partitions* a *shared business entity state* is located. It denotes that a *business entity* of type *registration* is in state *requested* after the *get customer ID* activity has finished. This object is in turn input to the act on *registration request* activity of the *selling organization*.

The register customer process (Figure 7–9) explained in detail

The *shared business entity state* denotes that the *registration business entity* in state *requested* is of relevance to both parties. After the *act on registration request* action there is a decision if a *credit check* is required or not. If the check is necessary, the *registration* object is now in state *pending credit check* denoted as *internal business entity state*. It is still the same *business entity* instance (as denoted by the *shared business entity state* before), even though it is now in another state and only of relevance to the *selling organization*. It follows, that *shared business entity states* depict business entity states that are of relevance to two partners, in contrary to *internal business entity states* identifying business entity states that are just of relevance to a partner's internal process.

Depending on certain conditions, there may be more than one successor of a *business entity state* or a *business process activity*. Decisions in the *business process activity model* are modeled by common UML means using *decision* nodes and *condition guards* on *transitions*. Concurrences in a process description are denoted by UML *forks* and *joins* ('synchronization bars').

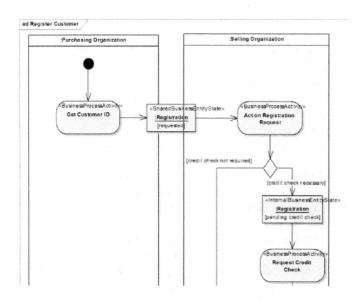

Fig. 7–9 Cut-out from the business process model specifying the register customer process from an interorganizational point of view.

7.2.1.1   Artifacts

The results of the *business process view* are **business process activity models** that refine the workflow of relevant business processes. UN/CEFACT recommends to refine every business process that might be of interest to further collaborations by a *business process activity model*. By using the means of the *business process view*, complexity in further modeling work steps is reduced.

7.2.2   Business Entity View

7.2.2.1   Overview and purpose

The *business entity view* describes relevant business entities that are affected by the execution of a business process. A *business entity* is a real-world thing that is of relevance to one or more business partners in a business process (e.g. "order", "account", etc.). Affecting a business entity usually implies changing its internal state. *Business entity life cycles* describe the flow of possible states of a *business entity*.

It is the goal of the *business entity view* to gather knowledge of items with business significance including the states they might adopt. Similar to the *business process view*, the *business entity view* is not a mandatory part of the *business requirements view*. However, UN/CEFACT suggests to capture all relevant business entities together with their lifecycle.

The business process view describes in detail real-world things having business significance

Business entity states are of further relevance to the *business process view* and the *business choreography view*. The *business process view* shows how *business entities* are affected by the execution of a business process. In other words, we detail how several steps in a process trigger changes of *business entity states* (See "Business Process View" on page 106). In the context of the *business choreography view*, we use *business entity states* to specify *transition guards* in *business collaboration protocols*. Guarding a *transition* by a *business entity state* implies that a *transition* is only effective if the respective *business entity* is in the corresponding state (See "Business Choreography View" on page 135).

A business entity states are relevant to the business process view and to the business choreography view

7.2.2.2   Stereotypes

BusinessEntity (Class): A *business entity* is a real-world thing having business significance for one or more business partners in a business process (e.g. "order", "account", etc.).

BusinessEntityState (State): A *business entity state* is a certain state that a *business entity* obtains in its lifecycle (e.g. an order is submitted, cancelled, accepted, etc.).

BusinessEntityLifecycle (StateMachine): A *business entity lifecycle* describes the flow in which *business entity states* occur.

7.2.2.3   Worksheets

Business entity view worksheet

Common Information
Business library information

Business entity worksheet

Common Information
Definition: A high-level definition describing the type of a *business entity* as well as its purpose.
Pre-conditions: Describes requirements that have to be fulfilled in order to execute a *business entity lifecycle*

Post-conditions: Identifies conditions that have to be satisfied just after the end of a *business entity lifecycle* (e.g. an order is confirmed).

Begins when: Specifies an event that initiates the lifecycle of a *business entity* - usually it creates the object (e.g. the buyer creates an order as soon as he has received a quote)

Ends when: Specifies an event that indicates the end of the *business entity lifecycle* in the context of the considered collaboration. This is mostly the last state in the lifecycle (e.g. an order is confirmed).

Exceptions: Identifies errors that may occur in the flow of a *business entity lifecycle*

Identify each state in the lifecycle of a business entity and describe it using the information defined beneath.

> *Common Information*
>
> *Definition:* A high-level definition describing the *business entity state* in detail
>
> *Predecessing states:* Identifies one or more predecessors of this state
>
> *Valid actions:* If a *business entity* reaches a certain state a set of business actions can be taken (e.g. once an order is in state accepted the notification of shipment of the order is a valid business action).

7.2.2.4 Step by step modeling guide

1. Identify *business entities*
2. Describe the lifecycle of each *business entity*

Identify business entities

The *business entity view* is one of the simplest and fastest steps in the UMM workflow. You may either describe all *business entities* in one *business entity view* or you split them up in multiple *packages*.

Business entities are usually identified in the *business process view*. In the *business entity view* each identified *business entity* is then described in detail by its own worksheet. Table 7–4 shows the worksheet for the *business entity order* that is part of the *order from quote* example.

Relevant business entities are identified in the business process view

Furthermore in our example, we define each *business entity* in its own *business entity view* (Figure 7–10). A *business entity lifecycle* describes in turn the states of a *business entity*. The *business entity lifecycle* is a stereotyped *state machine* and modeled as a child of the respective *business entity*. Regarding the *order from quote* example, Figure 7–10 shows the contents of the *registration* and the *credit check business entity lifecycle*.

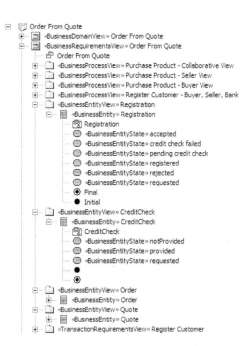

Fig. 7–10 Order from quote example: business entity view package structure

Describe the lifecycle of each business entity

A *business process entity lifecycle* contains a flow of states of a *business entity*. Define the sequence in which the *states* occur by connecting them via *transitions*. You may use UML *pseudo states* to denote *concurrences* and *decisions*.

Use business entity states to describe the business entity lifecycle

Describing a business entity's lifecycle is again based on the input captured by the corresponding worksheet. Regarding the *order from quote* example we derive the *business entity lifecycle* as shown in Figure 7–11 from the *order business entity* worksheet (Table 7–4). As the first one in the *order* lifecycle, the *business entity state submitted* is identified. *Submitted* is in turn the predecessing state for the *business entity states accepted* and *rejected*. In other words, after an *order* is *submitted* it may either be *accepted* or *rejected*. After an *order* is either *accepted* or *rejected* the end of the *business entity lifecycle* is reached.

Tab. 7–4Example worksheet for the order business entity

Form: BusinessEntity	
General	
Business Entity Name	Order
Definition	An order captures the goods or services that party A wants to buy from party B.
Description	An order is mostly a document that contains information about goods and services a business partner wants to order from another party. The respective goods and services that are demanded are itemized using line items. The acceptance of an order results in a residual obligation between the two parties (buyer and seller) to fulfil the terms of a contract.
Lifecycle	
Pre-condition	a quote is required
Post-condition	none
Begins When	an order is created by the buyer as soon as he receives a quote that is acceptable for him
Ends When	the order is confirmed (and filed correspondingly)
Exceptions	none
Lifecycle State (submitted)	
Name	submitted
Definition	The order is sent from the buying party to the selling party
Description	The buyer created the order and transmitted it to selling party. The selling party has then to decide about the acceptance of the order
Predecessing State	None (first state in the lifecycle)
Valid Actions	None identified
Lifecycle State (accepted)	
Name	accepted
Definition	The order is processed by the seller and he accepted the fulfilment of the order
Description	The seller accepts the order when he wants and is able to deliver the demanded goods/services. As soon as the seller accepts the order, an agreement is concluded between the two parties
Predecessing State	submitted
Valid Actions	Sending a notification of shipment
Lifecycle State (rejected)	
Name	rejected
Definition	An order is rejected if the seller refuses to fulfil the purchase order request issued by the buyer
Description	The seller is allowed to refuse an order request without assigning a reason for it. An order might be rejected if the seller is not able to meet the demands of the buyer due to various reasons.(e.g. the demanded goods are out of stock, requested allowances are unacceptable,…)
Predecessing State	submitted
Valid Actions	The seller might send a counter offer

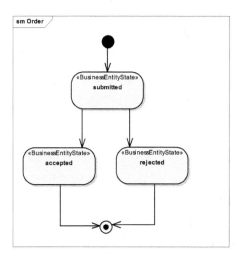

Fig. 7–11 *Order from*
quote example: the
business entity lifecycle of
the order business entity

7.2.2.1   Artifacts

Results from the *business entity view* are **business entities** that are of
relevance to further collaborations. For each *business entity* its **business
entity lifecycle** is described by a flow of *business entity states*.

7.2.3   Collaboration Requirements View

7.2.3.1   Overview and purpose

The preceding work steps determine the need for possible collaborations. In the *collaboration requirements view* we detail the requirements of
each identified collaboration. Requirements are again gathered by corresponding worksheets and depicted through *use case* descriptions. The
requirements we capture include a rough list of performed actions, start and
end characteristics (e.g. *begins when, ends when,* etc.) and the participants
of the respective collaboration.

A collaborative process is called a *binary collaboration* if exactly two
roles participate. If more roles are involved it is called a *multiparty collaboration.* The execution of a collaborative process includes several interactions between its participants. In other words, a business collaboration is
built up by one or more interactions between its participating roles. An
interaction may either be a business transaction or a nested business collab-

*In the collaboration
requirements view the participants of a collaboration
are determined*

oration. The high-level actions list captured in the process' requirements serves as a starting point to identify required interactions.

Furthermore the *collaboration requirements view* is used to map roles participating in a collaboration to the corresponding transaction roles or nested collaboration roles. In other words, this mapping defines which role of a collaboration performs a certain role in a transaction or in a nested collaboration. Business collaborations that are later constructed in the *business choreography view* rely on this mapping in order to collate actions to participating parties.

Collaboration roles are mapped to transaction roles

7.2.3.2 Stereotypes

BusinessCollaborationUseCase (UseCase): A *business collaboration use case* captures the requirements on a specific business collaboration between two or more involved partners. Business partners participate in a collaboration by playing an *authorized role*. A *business collaboration use case* is composed of *business transaction use cases* and other *business collaboration use cases*.

AuthorizedRole (Actor): An *authorized role* takes an active part in a collaboration or transaction. An *authorized role* is not a business partner, but an abstract concept. In the *collaboration realization view* concrete *business partners* are mapped to abstract *authorized roles*.

participates (Association): A *participates* association denotes that a certain *authorized role* takes part in the execution of a transaction or collaboration.

mapsTo (Dependency): Authorized roles of a *business collaboration use case* are mapped to *authorized roles* of included *business collaboration uses cases* or *business transaction use cases* via a *mapsTo* dependency.

7.2.3.3 Worksheets

Collaboration requirements view worksheet

Common Information
Business library information

Business collaboration use case worksheet

Common Information
Definition: Describes the overall customer value that is created by the business collaboration for all participants.

Participating roles: Identifies all roles participating in the business collaboration. Each identified participant is modeled as an *authorized role* in a further step and connected with the *business collaboration use case* via a *participates* association.

Affected business entities: Identifies *business entities*, which are affected by the execution of the business collaboration. In other words, this covers *business entities* whose states are changed during the execution of the business collaboration.

Pre-conditions: Describes requirements that have to be fulfilled in order to execute the business collaboration (e.g. a customer has to be registered to request a quote or to order something).

Post-conditions: Identifies conditions that have to be satisfied just after the execution of the business collaboration (e.g. an order is confirmed).

Begins when: Specifies business events that are required to initiate the business collaboration. (e.g. the order collaboration is started as soon as the buyer sends a quote request).

Ends when: Specifies business events that indicate the termination of a business collaboration (e.g. the order collaboration is finished when the contract between buyer and seller is established).

Exceptions: Identifies errors that may occur during the execution of a business collaboration. This listing covers errors that are not considered in the described collaboration flow.

Actions: Identifies all actions that are part of this business collaboration.

7.2.3.4 Step by step modeling guide

1. Describe the *business collaboration use case*
2. Identify the participants and denote them as *authorized roles*
3. Map collaboration roles to transaction roles or nested collaboration roles

The need for each identified collaboration is manifested by its own *business collaboration use case*. Each *business collaboration use case* has in turn to be modeled in its own *collaboration requirements view* together with its participating roles.

Describe the business collaboration use case

We start by modeling the *business collaboration use case* according to the corresponding worksheet information. The *business collaboration use case* worksheet captures the purpose and a listing of high-level actions of a *business collaboration use case*.

Considering the *order from quote* example, Table 7–5 shows the worksheet for the *order from quote* business collaboration. Based on this worksheet we model a corresponding *business collaboration use case* (Figure 7–12).

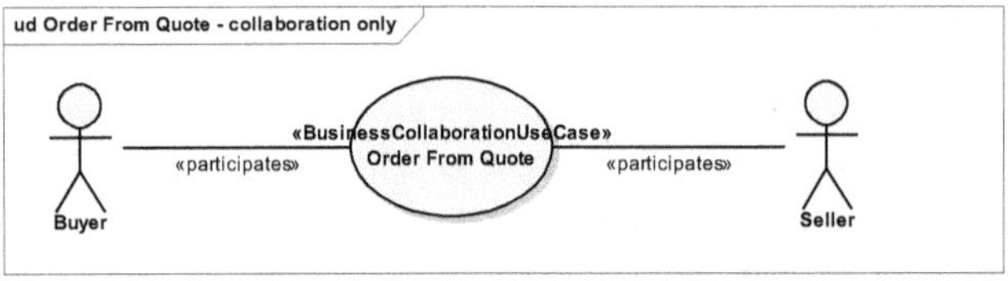

Example: order from quote

***Fig. 7–12** Business collaboration use case order from quote*

Identify the participants and denote them as authorized roles

The worksheet further identifies the participants of the collaborative business process. We add one *authorized role* for each participant to the *collaboration requirements view*. Then, each *authorized role* has to be connected with the *business collaboration use case* via a *participates* association. Depending if the collaboration is binary or multiparty, two or more *authorized roles* are involved.

Model the participating roles

***Tab. 7–5**Example worksheet for order from quote*

Form: BusinessCollaborationUseCase	
General	
Business Collaboration Name	Order From Quote
Definition	The purpose of this business collaboration is provide a means for a buyer to request a quote for required items from a seller and to provide a means for a seller to provide the buyer with a formal quote or quote rejection.
Description	Once the buyer has received the quote, the buyer may chose to purchase the items from the seller. If so, this business collaboration provides a means for the buyer to send to the seller a list of items that the buyer desires to purchase, and to provide a means for the seller to send to the buyer a formal acceptance or rejection of the buyer's order.
Start/End Characteristics	
Affected Business Entities	- Quote - Order
Pre-condition	Registration.registered
Post-condition	Order.accepted or Order.rejected
Begins When	Quote Requestor send to Quote Responder a list of items for quotation
Ends When	Buyer receives from a seller a formal acceptance or rejection of an order placed with the seller
Exceptions	None identifed
Actions	
Description	- The buyer sends to the seller a list of items that the buyer requests the seller to quote. - The seller sends to the buyer a formal quote or quote rejection - If the buyer chooses to purchase from the seller, the buyer will send an order to the seller. - The seller will send to the buyer a formal acceptance or rejection of the buyer's order.

An *authorized role* taking part in a certain collaboration must be defined in the same *collaboration requirements view* as the corresponding *business collaboration use case*. If an *authorized role* with the same name (e.g. buyer, payer, etc.) participates in multiple collaborations a different *authorized role* has to be defined for each *business collaboration use case* (thus in each *collaboration requirements view*). The *collaboration requirements view* represents a namespace, hence two equally named *authorized roles* that are modeled in different *collaboration requirements views* are not identical. It follows, that two *authorized roles* in the same *collaboration requirements view* must not have an identical name. The *transaction requirements view* and the *collaboration realization view* use the same mechanism to handle *authorized roles*.

A collaboration requirements view is a namespace for the elements it contains

In our example we add one *authorized role* named *buyer* and one named *seller* to the corresponding *collaboration requirements view*. Then connect both with the *business collaboration use case order from quote* using a *participates* relation (Figure 7–12).

Order from quote example: add authorized roles for buyer and seller

Map collaboration roles to transaction roles or nested collaboration roles

In the last stage of the *collaboration requirements view* we have to determine which collaboration participant plays which role in an included business transaction or nested business collaboration. Similar to *business collaboration use cases*, *business transaction use cases* involve participating roles. However, the number of participating roles is limited to exactly two roles. *Business transaction use cases* are further detailed in chapter 7.2.4.

Switching between collaboration requirements view and transaction requirements view is required

Once all included *use cases* and their participating roles are fixed, a mapping of roles to the outer collaboration must take place. Each role of an included transaction or nested collaboration has to be connected to exactly one role of the outer collaboration. We denote this relationship via a *mapsTo* dependency leading from the *authorized role* of the outer *business collaboration use case* to the *authorized role* of the *business transaction use case* or nested *business collaboration use case*.

Mapping collaboration roles to transaction roles and inner collaboration roles

These mapping definitions have some logical constraints: only one role of a *business collaboration use case* must have a *mapsTo* dependency to a certain role of an included *business transaction use case* or *business collaboration use case*. It follows, that each role of an included *business transaction use case* or *business collaboration use case* must be connected to exactly one role of the including *business collaboration use case*.

Some constraints on relationships between participants

We demonstrate this modeling step again with our *order from quote* example. Figure 7–13 shows the mapping between roles participating in the *order from quote* collaboration and the particular roles of the *request for quote* and *place order* transaction.

Order from quote: mapping roles of the order from quote collaboration

Fig. 7–13 *Order from quote example: mapping collaboration roles to transaction roles and nested collaboration roles*

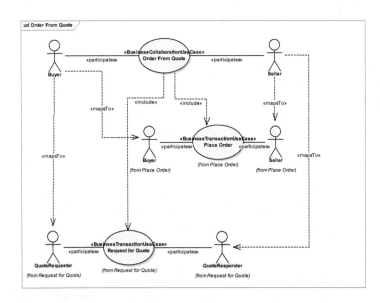

It is important that the *business collaboration use case* and the two participating roles *buyer* and *seller* are the same as those shown in Figure 7–12. It is just another diagram which exposes a different view on the same model elements. In other words, the relationships that are defined in Figure 7–13 apply also to the *business collaboration use case* in Figure 7–12 after they have been defined in some other diagram (in our example Figure 7–13). However, if we look again at Figure 7–12, we would not see these relationships until we drag the associated elements (e.g. the *business transaction use case place order*) onto this diagram. Thus, we see that diagrams just expose different views on the same model.

Diagrams lie!

In our example we have two participants - *buyer* and *seller* - in the *order from quote* collaboration. These participants are denoted as *authorized roles* and linked with the *business collaboration use case* via a *participates* relationship. The *order from quote* collaboration comprises two transactions - *request for quote* and *place order*, which are denoted by *business transaction use cases*. In order to indicate that the business transactions are executed within the *order from quote* collaboration we connect them via an *include* relation. Such an *include* relation leads from the *business collaboration use case* to the *business transaction use cases*.

Considering the order from quote collaboration

A *business transaction* is always conducted between exactly two roles. Thus, a *business transaction use case* must always be connected with two *authorized roles*. The *business transaction use case request for quote* is

Examining request for quote

associated with the *authorized roles quote requestor* and *quote responder*. All three model elements are annotated with *from request for quote*. This means that these elements are part of a different *package* (but visualized on this diagram that is located in the *order from quote collaboration requirements view*) and hence appear in a different namespace.

Considering the *business transaction use case place order* we have again two associated *authorized roles* - *buyer* and *seller*. Again the namespace annotation *from place order* points out that the *business transaction use case* and the two *authorized roles* are part of a different *package*. Furthermore the namespace clarifies that the *authorized roles* connected with the *business collaboration use case* are not the same as the two associated with the *business transaction use case*, although they are named identically.

The two buyers are not identical - the place order transaction and its participants

In the last step we determine which *authorized role* participating in the *order from quote* collaboration plays which role of the two included *business transaction use cases*. We indicate that the *buyer* of the *order from quote* collaboration takes up the role of the *quote requestor* in the *request quote* transaction using a *mapsTo* association. Similarly, in order to specify that the *seller* (of the *order from quote* collaboration) plays the role of the *quote responder*, we drag a *mapsTo* from the *seller* to the *quote responder*. It follows, that a *mapsTo* leads always from the role participating in the collaboration to the respective role of the transaction or nested collaboration. Considering the *place order* transaction we want to define that the buying role is fulfilled by the *buyer* of the *order from quote* collaboration. Thus, the *seller* of the collaboration is specified to be the *seller* in the *place order* interaction. In order to manifest this mapping we connect the *buyer* of the *order from quote* collaboration with the *buyer* of the *place order* transaction and similarly the *seller* of the collaboration with the *seller* of the transaction.

The buyer of the collaboration maps to the quote requestor of the place order transaction etc....

It is not necessary to depict *mapsTo* relationships in use case diagrams. You might also specify these relationship directly to the model elements without visualizing them on diagrams. Towards a straightforward mapping of roles, adopting *use case diagrams* is recommended though.

7.2.3.1 Artifacts

The *collaboration requirements view* results in a **business collaboration use case** capturing the requirements on a collaboration. Participants of a collaboration are identified and described as **authorized roles**. Furthermore, relations to *authorized roles* participating in included transactions or collaborations are specified.

7.2.4 Transaction Requirements View

7.2.4.1 Overview and purpose

The purpose of the *transaction requirements view* is quite similar to the *collaboration requirements view*. However, this step focuses on atomic interactions between partners in the workflow of a collaboration, so called transactions. Transactions describe an information exchange between roles, consisting of a request and an optional response.

Interactions are performed between exactly two participants. Thus, exactly two *authorized roles* take part in a *business transaction use case*. It is a task of this view to describe the two *authorized roles* participating in the transaction.

Transaction describe an information exchange between two roles

7.2.4.2 Stereotypes

- *BusinessTransactionUseCase (UseCase):* A *business transaction use case* describes the requirements of a business transaction. A business transaction describes an atomic one-way or two-way message exchange between exactly two participants.
- *AuthorizedRole (Actor):* An *authorized role* takes an active part in a collaboration or transaction. An *authorized role* is not a business partner, but an abstract concept. Later concrete *business partners* are mapped to collaboration roles which in turn map to transaction roles.
- *participates (Association):* A *participates* association denotes that a certain *authorized role* takes part in the execution of a transaction or collaboration.

7.2.4.3 Worksheets

Transaction requirements view worksheet

- *Common Information*
- *Business library information*

Business transaction use case worksheet

- *Common Information*
- *Definition:* Describes the overall customer value that is created by the *business transaction* for all participants.
- *Requesting role:* Identifies the role that initiates the business transaction. The identified participant is then modeled as an *authorized role*

and connected with the *business transaction use case* via a *participates* association.

Responding role: Identifies the role that is the reactor of the business transaction. Using this information a new *authorized role* is modeled and connected with the *business transaction use case* via a *participates* association.

Requesting activity: Specifies the name of the activity performed by the requesting role in the business transaction

Responding activity: Specifies the name of the activity performed by the responding role in the business transaction.

Affected business entities: Identifies *business entities*, which are affected by the execution of the business transaction. In other words, this information covers business entities whose state is changed by executing the business transaction.

Pre-conditions: Describes requirements that have to be fulfilled in order to execute the business transaction (e.g. a quote has to be requested prior to an order submission).

Post-conditions: Identifies conditions that hold after the execution of the business transaction (e.g. a quote is issued).

Begins when: Specifies business events for the initiation of the business transaction. It may be used to specify a semantic state, that has to be reached in order to start the process (e.g. the place order transaction starts when the buyer identifies the need to fill up stock).

Ends when: Specifies business events that indicate the termination of a business transaction (e.g. the place order transaction is finished when the buyer receives the order confirmation from the seller).

Exceptions: Identifies errors that may occur during the execution of a business transaction.

7.2.4.4   Step by step modeling guide

1. Describe the *business transaction use case*
2. Identify participating roles

Similar to *business collaboration use cases*, a *business transaction use case* captures the requirements of exactly one *business transaction*. Hence, model each *business transaction use case* in its own *transaction requirements view*.

One business transaction use case per transaction requirements view

Tab. 7–6 *Order from quote example: worksheet capturing the requirements of the business transaction use case request for quote*

Form: BusinessTransactionUseCase	
General	
Business Transaction Name	Request for Quote
Definition	The purpose of this business transaction is to allow a means for a quote requestor to send a list of items to a quote responder for quotation, and to provide a means for the quote responder to send to the quote requestor either a formal quote or a quote rejection.
Requesting Role	QuoteRequestor
Responding Role	QuoteResponder
Requesting Activity	obtain quote
Responding Activity	calculate quote
Start/End Characteristics	
Affected Business Entities	Quote
Pre-condition	Registration.registered
Post-condition	Quote.provided or Quote.refused
Begins When	quote requestor sends quote to quote responder
Actions	obtain quote; calculate quote
Ends When	quote requestor receives from quote responder a quote or quote rejection
Exceptions	none identified

Tab. 7–7 *Order from quote example: worksheet for the business transaction use case place order*

Form: BusinessTransactionUseCase	
General	
Business Transaction Name	Place Order
Definition	The purpose of this business transaction is to provide a means for a buyer to send to the seller a list of items that they wish to purchase, and to provide a means for a seller to send to the buyer a formal acceptance or rejection of the buyer's order.
Requesting Role	Buyer
Responding Role	Seller
Requesting Activity	submit order
Responding Activity	process order
Start/End Characteristics	
Affected Business Entities	Order
Pre-condition	Quote.provided
Post-condition	Order.accepted or Order.rejected
Begins When	buyer sends order to seller
Actions	submit order; process order
Ends When	buyer receives from seller a formal acceptance or rejection of the order placed with the seller
Exceptions	none identified

Describe the business transaction use case

Modeling the *transaction requirements view* starts with the *business transaction use case*. Information detailing its requirements and its desired behavior is captured in the appropriate worksheet for a *business transaction use case.* In our *order from quote* example we derive a *business transaction use case* called *request for quote* (Figure 7–14) from the corresponding worksheet shown in Table 7–6. Similarly, we model a *business transaction use case* named *place order* (Figure 7–15) based on the respective worksheet input shown in Table 7–7.

Identify participating roles

As described above, exactly two *authorized roles* take an active part in a transaction. Thus, we add two *authorized roles* to the *transaction requirements view* and connect each of them with the *business transaction use case* via a *participates* association. We derive the names for the *participating authorized* roles from the corresponding worksheet fields (requesting and responding role). In our *request for quote* transaction (Figure 7–14) there is one role called *quote requestor* and one *quote responder.*

Derive the participating roles from the business transaction use case worksheet

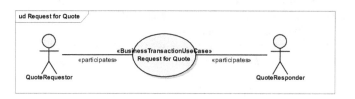

Fig. 7–14 Order from quote example: business transaction use case register customer

In a similar way, we derive the roles *buyer* and *seller* taking part in the *place order business transaction use case* (Figure 7–15). Again we indicate that both *authorized roles* take part in the *business transaction use case* by connecting them with the *business transaction use case* via a *participates* association.

Fig. 7–15 Order from quote example: business transaction use case place order

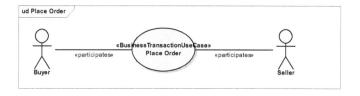

Similar to the *collaboration requirements view,* each role participating in a transaction must be defined in the particular *transaction requirements view.* If the same role (e.g. payer, seller, etc.) takes part in more than one transaction, add one *authorized role* with the same name to each *transaction requirements view.*

7.2.4.1 Artifacts

Results from the *transaction requirements view* are the requirements of identified business transactions described by **business transaction use cases**. We create *use case diagrams* containing the *business transaction use case* and its two participating *authorized roles*. Furthermore, the *business transaction use case* worksheet captures a set of requirements of the corresponding business transaction.

7.2.5 Collaboration Realization View

7.2.5.1 Overview and purpose

The *collaboration realization view* is a concept to assign *business partners* to *authorized roles* participating in a collaboration. In the UMM a direct association of *business partners* with collaboration or transaction roles is not allowed. *Collaboration realization views* allow that different sets of *business partners* perform the same collaboration.

A business collaboration realization realizes a business collaboration with a specific set of business partners

The roles participating in a *business collaboration realization* correspond to the roles of the *business collaboration use case* that is realized. *Business partners* are mapped to collaboration realization roles in order to define which role a business partner plays in a certain collaboration. A *business collaboration realization* captures no additional requirements, but the requirements of the realized collaboration apply also for the collaboration realization.

Business partners take up roles of a collaboration realization

The *collaboration realization view* is a highly scalable concept that allows that the same collaboration is performed by an unlimited set of business partners. Thus, it boosts reusability of identified business collaborations with no increase in complexity.

The collaboration realization view enhances scalability

7.2.5.2 Stereotypes

BusinessCollaborationRealization (UseCase): A *business collaboration realization* realizes an abstract collaboration between a set of *business partners*. *Business collaboration realizations* capture no additional requirements, but the requirements of a realized collaboration also apply for the collaboration realization.

AuthorizedRole (Actor): An *authorized role* takes an active part in a collaboration or transaction. *Authorized roles* of a *business collaboration realization* correspond to the *authorized roles* of the realized *business collaboration use case*. Thus no additional semantic is added. *Business partners* are mapped to *authorized roles*, to indicate which role a partner performs in a certain collaboration.

participates (Association): A *participates* association denotes that a certain *authorized role* takes part in a collaboration realization.

mapsTo (Dependency): Authorized roles of a collaboration realization are mapped to their corresponding collaboration roles. Finally, the *mapsTo* relation is used to indicate which *authorized role* of a *business collaboration realization* a *business partner* has.

BusinessPartner (Actor): A *business partner* is a certain type of party taking up a role in the execution of a collaboration. *Business partners* are not defined in the *collaboration realization view*, but used during the modeling workflow in order to indicate which role is performed by a certain partner in a collaboration.

7.2.5.3 Worksheets

Collaboration realization view worksheet

Common Information
Business library information

Business collaboration realization worksheet

Common Information
Business collaboration specification: Identifies the business collaboration, which is realized by this *business collaboration realization*.
Participating roles: Denotes the *authorized roles* that participate in the *business collaboration realization*. Usually, these roles correspond to the roles of the business collaboration.
Partner roles: Identifies business partners that participate in the collaboration. Furthermore, it is specified which role is played by which business partner.

7.2.5.4 Step by step modeling guide

1. Describe the *business collaboration realization* and identify the realized business collaboration
2. Map *business partners* to *authorized roles*

Modeling *collaboration realization views* is the last step in the BRV. Each *business collaboration realization* is described in its own *collaboration realization view.* Thus, model exactly one *business collaboration realization use case* in one *collaboration realization view.* In order to facilitate modeling in this view using a *use case diagram* is recommended. Model elements that are not part of the *collaboration realization view,* but used in this view (e.g. *business collaboration use case, business partners*) may be dragged on the *use case diagram.*

Describe each business collaboration realization in its own collaboration realization view

Tab. 7–8 Order from quote example: worksheet for the order from quote collaboration realization

Form: BusinessCollaborationRealization	
General	
Realization Name	Order From Quote
Business Collaboration Specification	Order From Quote
Participants	Buyer, Seller
Partner roles	Purchasing Organization (Buyer), Selling Organization (Seller)

Describe the business collaboration realization and identify the realized business collaboration

Add a *business collaboration realization* to the *collaboration realization view.* It is suggested, that you name the collaboration and its realization equally. In order to indicate that a *business collaboration realization* realizes a certain *business collaboration use case* connect them with a *realize* dependency. The *realize* dependency leads from the *business collaboration realization* to the *business collaboration use case.*

In our *order from quote* example we specify a *business collaboration realization* named *order from quote* (see Figure 7–16) based on the worksheet input of Table 7–8. Then we drag a *realize* association from the *business collaboration realization* to the *business collaboration use case.* We facilitate this by dragging the already existing *business collaboration use case* onto the use case diagram.

Order from quote example

Add one *authorized role* to the *collaboration realization view* for each participating *authorized role* of the realized *business collaboration use case.* Connect the *authorized roles* of the *business collaboration realization use case* with their corresponding *authorized roles* of the *business collaboration use case* via a *mapsTo* dependency. The *mapsTo* dependency leads from the collaboration realization roles to the business collaboration roles. Exactly one *authorized role* of the *business collaboration realization use case* must be connected with exactly one *authorized role* of the *business collaboration use case.* In order to reduce complexity we suggest to use the same names for the *authorized roles* of the *collaboration requirements view* and the *collaboration realization view.*

Each role of a collaboration realization corresponds to one role of a business collaboration use case

Considering Figure 7–16 we add two new *authorized roles* to the *collaboration realization view* - one called *buyer* and one called *seller*. As recommended we name these roles identically to the participants of the *business collaboration use case*. Subsequently, we denote the relationship between the *buyer* of the *business collaboration realization* and the *buyer* of the *business collaboration use case* using a *mapsTo* association. The *seller-to-seller* relationship is similarly constituted via a *mapsTo*.

Fig. 7–16 Business collaboration realization order from quote and the assignment of business partners to collaboration realization roles

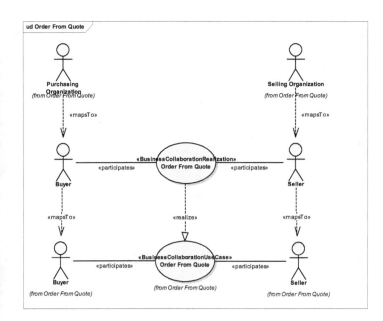

Map business partners to authorized roles

In a last step, define which business partner takes up a certain role in a collaboration realization. Therefore, connect a *business partner* with the corresponding *authorized role* of the *business collaboration realization* via a *mapsTo* dependency leading from the *business partner* to the *authorized role*.

Two *business partners* - a *purchasing organization* and a *selling organization* - are taking up an active part in our *order from quote* example. The *purchasing organization* plays the role of the *buyer* in the *order from quote* collaboration. Hence, we indicate this participation via *mapsTo* association leading from the *purchasing organization* - the *business partner* - to the *buyer* - the *authorized role*. We do the same for the *business partner* named

Define the business partner that plays a certain role

Order from quote example: the purchasing organization plays the buyer role...

selling organization in order to specify its commitment to fulfill the selling part in the *order from quote* collaboration.

Each *business partner* performs at most one role in one *collaboration realization view*, but the same *business partner* may play different *authorized roles* in different *collaboration realization views*.

7.2.5.1   Artifacts

The *collaboration realization view* results in a formal *use case description*. Amongst other things it denotes which business partner plays which role in the collaboration under consideration. Each different set of business partners performing a collaboration results in one **business collaboration realization** of that collaboration, hence also in one *collaboration realization view*.

7.3   Business Transaction View

7.3.0.1   Overview and purpose

The *business transaction view* is the last of the three main views in the UMM workflow. In this stage business collaborations are constructed in accordance to the requirements collected in previous work steps. Artifacts in this view are modeled from the business analyst's point of view, based upon the knowledge and artifacts gained from the business requirements view.

Business collaborations are constructed in the business transaction view

A business collaboration is composed of interactions between its participating partners. It is the purpose of the *business transaction view* to describe the choreography of interactions and the information that is exchanged in these interactions.

Business collaborations are composed of interactions (transactions) between roles

Modeling the *business transaction view* is divided in three subviews that together describe the overall choreography of information exchanges [FOU03]. The *business choreography view* defines the flow of business collaborations which may be composed of several interactions. Business collaborations are described by means of *business collaboration protocols* and implement the requirements gathered from the corresponding *collaboration requirements views*.

Interactions between business partners result in an information exchange that consists of a request and an optional response. The concept of a *business transaction* is used in UMM to capture an interaction between participants. *Business transactions* are described in *business interaction views*, the requirements on *business transactions* are captured in the corresponding *transaction requirements views*.

Interactions are described by business transactions, their requirements are captured in transaction requirements views

The information that is exchanged in a *business transaction* is captured   *Exchanged information is*
in the third subview of the *business transaction view*. *Business information*   *modeled in business infor-*
views are used as containers to model the structure of exchanged informa-   *mation views*
tion.

The *business transaction view* covers all aspects to construct executable
B2B collaborations between the information systems of business partners.
Business collaboration models developed in this stage may be transformed
to machine-interpretable process descriptions in a further step. Chapter 9
explains how a BPEL process description is derived from a UMM business
collaboration.

7.3.0.2   Stereotypes

- *BusinessChoreographyView (Package):* A *business choreography view*
  is a container to capture all artifacts describing the flow of a business
  collaboration.
- *BusinessInteractionView (Package):* A *business interaction view* is a
  container to capture all artifacts describing an interaction between two
  business partners.
- *BusinessInformationView (Package):* A *business information view* is a
  container to capture the structure of information that is exchanged in an
  interaction.

7.3.0.3   Worksheets

The worksheets of the *business transaction view* are described in the continuative sections of the particular subviews of which they are part of. However, we utilize the *business library* worksheet to gather common information about the *business transaction view package*.

Business transaction view worksheet

- *Common Information*
- *Business library information*

7.3.0.4   Step by step modeling guide

1. Choreograph business collaborations by means of *business collaboration protocols*
2. Describe information exchanges using *business transactions*
3. Specify information that is exchanged in *business transactions*

The *business transaction view* is a container for the three subviews: *business choreography view, business interaction view* and *business information view*. It must not contain any other modeling elements.

Choreograph business collaborations by means of business collaboration protocols

Modeling the *business transaction view* should be started with describing the flow of business collaborations. A business collaboration is described by a *business collaboration protocol*. Each *business collaboration protocol* needs to be placed in its own *business choreography view*. Thus add a *business choreography view* for every business collaboration that has to be defined.

Each business collabora-tion protocol must be speci-fied in its own business choreography view

Describe information exchanges using business transactions

A business collaboration is composed of one to many information exchanges between its participants. In this version of the UMM, *business transactions* are the only concept to describe these interactions. Each *business transaction* has to be modeled in its own *business interaction view*.

Place each business trans-action in its own business interaction view

Specify information that is exchanged in business transactions

In order to model the structure of the business information that is exchanged in *business transactions* you may use one or more *business information views*. UMM makes no restriction here: you may utilize multiple *business information views* to gain a semantic separation or you may use only one *business information view*.

Describe business informa-tion in business information views

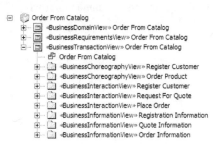

Fig. 7–17 Order from quote example: package structure in the BTV

Modeling the *business transaction view* in our *order from quote* example results in the *package* structure shown in Figure 7–17. We choreograph two business collaborations - *register customer* and *order from quote* - each in its own *business choreography view*.

Explaining the BTV pack-age structure of the order from quote example

The *business collaboration protocols* include a set of *business transactions*. Each *business transaction - register customer, request for quote* and *place order* - is specified in detail in a *business interaction view*.

Finally, we describe the information that is exchanged in *business transactions*. In our example, each *information envelope* is modeled in its own *business information view*. This results in three *business information views*. In general, it is up to the modeler to decide upon the number of *business information views*.

7.3.0.1 Artifacts

Modeling the *business transaction view* results in **business collaboration protocols**. Defining a choreography, the interactions that build up a collaboration are specified by **business transactions**. The **business information that is exchanged** in *business transactions* is captured using class descriptions. Artifacts that are derived in the *business transaction view* may be used in further steps to generate machine-executable process specifications.

7.3.1 Business Choreography View

7.3.1.1 Overview and purpose

In a *business choreography view* the workflow of a business collaboration is described. Business collaborations cover one or more interactions between their participating partners. The order of occurrence of these interactions might be complex including concurrences and decision nodes that influence the execution order. Choreographies are used to indicate possible flows in the execution order of interactions that are part of a collaboration. In order to query or reference the state of a business collaboration, the concept of persistent representations are used to capture the currently performed step in the execution workflow.

The business choreography view focuses on the flow of complex business collaborations

In UMM choreographies are defined by the abstract concept of a *business choreography behavior*. A *business choreography* is a persistent representation of a *business choreography behavior* that captures the current state of its execution. In this version of the UMM *business collaboration protocols* are the only valid specialization of a *business choreography behavior* to describe the flow of a collaboration. In future versions other concepts may be developed to describe the flow of collaborative processes.

An interaction in a *business collaboration protocol* may be either denoted as a *business transaction activity* or a *business collaboration activity*. *Business transaction activities* reference to an underlying *business transaction*.

Business transaction activities are refined by business transactions

A *Business collaboration activity* is refined by another *business collab-* *Business collaboration pro-*
oration protocol. Thus, *business collaboration protocols* may be nested to   *tocols may be nested using*
model complex collaborations in more detail. *Business collaboration proto-* *business collaboration*
cols are reusable artifacts, hence one *business collaboration protocol* may *activities*
refine more than one *business collaboration activity*.

A *business collaboration protocol* complies to the requirements speci- *Roles that participate in a*
fied in a corresponding *business collaboration use case*. This includes the *business collaboration are*
definition of roles that participate in a certain collaboration. In order to indi- *specified in the collabora-*
cate which role of a *business transaction* or nested *business collaboration* *tion requirements view*
protocol is played by which role of the (outer) *business collaboration proto-*
col, a well-defined role mapping mechanism is applied in the workflow of
the *collaboration requirements view* (see 7.2.3). Two or more roles may be
actively involved in a collaboration. If exactly two roles participate, a collaboration is called a binary collaboration, otherwise if more than two roles
participate, it is called a multiparty collaboration.

7.3.1.2   Stereotypes

BusinessChoreography (Class): A *business choreography* is a persistent
representation of the current state of execution of a business collaboration. In this version of the UMM, the behavior of a *business choreogra-*
phy must be described by a *business collaboration protocol*.

BusinessCollaborationProtocol (ActivityGraph): A *business collabora-*
tion protocol choreographs the flow of activities of a business collaboration. Activities of a *business collaboration protocol* may either be *busi-*
ness transaction activities or *business collaboration activities*.

BusinessTransactionActivity (ActionState): A *business transaction*
activity is an activity within a *business collaboration protocol* and indicates an information exchange between exactly two roles. A *business*
transaction activity is refined by a *business transaction*.

BusinessCollaborationActivity (ActionState): A *business collaboration*
activity is an activity within a *business collaboration protocol* and indicates a complex interaction between two or more collaboration roles. A
business collaboration activity is refined by a *business collaboration*
protocol. Thus, *business collaboration protocols* may be nested recursively.

mapsTo (Dependency): A *mapsTo* dependency is used in the *business*
choreography view to specify the refining *business transaction* of a
business transaction activity or the refining *business collaboration pro-*
tocol of a *business collaboration activity*. In both cases the *mapsTo*
leads from the *activity* to the *activity graph*. Furthermore a *mapsTo*
dependency is needed to depict which *business collaboration use case*

holds the requirements of a *business collaboration protocol*. In order to indicate this relation add a *mapsTo* leading from the *business collaboration protocol* to the *business collaboration use case*.

7.3.1.3 Worksheets

Business collaboration view worksheet

Common Information
Business library information

Business collaboration worksheet

Common Information
Definition: Describes the overall customer value that is created by the business collaboration for all participants.

7.3.1.4 Step by step modeling guide

1. Model the *business choreography* and the *business collaboration protocol*
2. Associate the *business collaboration protocol* with the corresponding *business collaboration use case*
3. Describe the choreography of the *business collaboration protocol*

Constructing business collaborations is the major goal of the UMM. This stage is one of the most important in the modeling workflow. In this version of the UMM, a *business collaboration protocol* is the only valid concept to choreograph the flow of a business collaboration. Each *business collaboration protocol* must be described in its own *business collaboration view*.

Tab. 7–9 Worksheet describing the order from quote choreography

Form: BusinessChoreography	
General	
Name	Order From Quote
Definition	The purpose of this business collaboration is provide a means for a buyer to request a quote for required items from a seller and to provide a means for a seller to provide the buyer with a formal quote or quote rejection.
Description	Once the buyer has received the quote, the buyer may chose to purchase the items from the seller. If so, this business collaboration provides a means for the buyer to send to the seller a list of items that the buyer desires to purchase, and to provide a means for the seller to send to the buyer a formal acceptance or rejection of the buyer's order.

Model the business choreography and the business collaboration protocol

At first, the persistent representation of a *business collaboration proto-col* needs to be described by a *business choreography*. Thus add one *business choreography class* to the *business choreography view*. A *business collaboration protocol* specifies the behavior of a *business choreography*. Hence, add the *business collaboration protocol* as a child of the *business choreography*. This indicates that the *business choreography* always captures the current state of the *business collaboration protocol* execution. Considering Figure 7–18 we explain this structure again by the *order from quote* collaboration. A *business choreography* (BC) named *order from quote* is added to a *business choreography view* (also called *order from quote*). The *business choreography* captures the state of the *order from quote business collaboration protocol* (BCP), which is modeled as a child of the *business choreography*.

Specify the business chore-ography that captures the current state in the execu-tion of a business collabo-ration protocol

Fig. 7–18 Relationship between a business choreography (BC) and a business collaboration protocol (BCP)

Associate the business collaboration protocol with the corresponding business collaboration use case

Afterwards, associate the *business collaboration protocol* with the cor-responding *business collaboration use case* that captures its requirements. In order to indicate this relation a *mapsTo* dependency is used, leading from the *business collaboration protocol* to the *business collaboration use case*.

A business collaboration protocol maps to a business collaboration use case

Describe the choreography of the business collaboration protocol

Table 7–9 shows the example worksheet for the *order from quote busi-ness choreography*. The choreography worksheet captures only basic input for the upcoming collaboration (e.g. name, definition,...). The main task - describing a *business collaboration protocol* as a flow of interactions between its participants - is accomplished by the business analyst using the worksheet input from the corresponding *business collaboration use case*. The *business collaboration use case* worksheet further references the actions (transactions and nested collaborations) that are performed during

The choreography of a business collaboration pro-tocol is constructed using the worksheet input from the corresponding business collaboration use case

its execution. The business analyst constructs the choreography of the collaboration, using the start and end characteristics of these actions, which are described again in their particular worksheets.

The main task of the *business choreography view* describes the choreography of the collaboration by an *activity graph* showing *activities* as well as UML *pseudo* and *final states*. *Activities* used in a *business collaboration protocol* may be *business transaction activities* and *business collaboration activities*. A *business transaction activity* indicates a simple interaction between two partners that result in an information exchange. This information exchange is further described by a *business transaction*. Connect the *business transaction activity* with its refining *business transaction* via a *mapsTo* dependency, leading from the *business transaction activity* to the *business transaction*. In this step of the modeling workflow, *business transactions* have not been described yet. Thus you may define the *mapsTo* mapping after you have described all *business transactions* in the *business interaction view*. Another alternative is switching between *business choreography view* and *business interaction view* and modeling a refining *business transaction* immediately after the *business transaction activity*.

Business transaction activities denote a simple information exchange between two authorized roles

A *business transaction activity* might be executed more than once at the same time in order to serve collaborations with different partners. Furthermore, a *business transaction activity* must be finished within a given timeframe. These features are specified via the *is concurrent* and the *time to perform* properties.

Business collaboration activities indicate a complex interaction between two or more roles, hence a *business collaboration activity* is further described by another *business collaboration protocol*. Nested *business transactions* or *business collaboration protocols* may be reused to refine multiple *business transaction activities* or *business collaboration activities*. Modeling the relationship between a *business collaboration activity* and a refining *business collaboration protocol* is done via a *mapsTo* dependency. Similar to the relationship between *business transaction activity* and *business transaction*, the *mapsTo* leads from the *business collaboration activity* to the nested *business collaboration protocol*.

Use business collaboration activities to nest business collaboration protocols

In order to describe a complex control flow common UML *pseudo* and *final states* may be used. *Pseudo states* may include *initial states*, *decisions*, *forks* and *joins*. *Initial States* denote the start of a *business collaboration protocol*. The successor of an *initial state* should be a *business transaction activity* or a *business collaboration activity*, it is not recommended to use a *pseudo state* in this place. Furthermore, the *transition* between the *initial state* and the first *action state* should not have a *transition guard*. A *decision* indicates a choice between two or more possibilities using *condition guards*. Concurrences in a flow of *activities* are indicated via *forks* and *joins*. *Final*

Use transitions, pseudo and final states to choreograph the flow of execution

states signalize the end of a *business collaboration protocol*. UN/CEFACT
suggests, but does not mandate, to use at least one *final state* to indicate a
failure and one to indicate a successful execution. Similar as for *initial
states*, only *action states* should be predecessors of *final states*. As in common UML *activity graphs*, *transitions* are used to denote the flow between
action states of a *business collaboration protocol*.

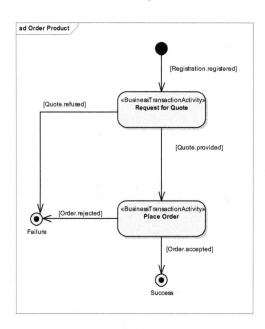

*Fig. 7–19 Example
business collaboration
protocol choreographing
the order from quote
collaboration*

If an *action state* - either a *business transaction activity* or a *business
collaboration activity* - or a *pseudo state* has more than two outgoing *transitions*, *transition guards* must be used. UN/CEFACT suggests to use *business entity states* - from the *business entity view* in the BRV - to guard *transitions*. This indicates that a *business entity* has to be in a certain state in
order to execute the *transition* (e.g. an order must be submitted). Another
alternative denoting *transition guards* are BPSS *condition expressions*
[BPS03]. Anyway, there is no strict schema that defines the notation of *transition guards*, but it is strongly suggested to use *business entity states* or
other semantically meaningful notations.

*Specifying transition
guards*

The *order from quote* collaboration results in the *business collaboration
protocol* shown in Figure 7–19. Starting from the *initial state* the first interaction is the *business transaction activity* named *request for quote*. This
business transaction activity is refined by an identically named *business
transaction* that details an information exchange. As the guard on the *tran-*

*Example: examining the
order from quote collaboration in detail*

sition leading to *request for quote* indicates, the *business entity registration* must be in state *registered* in order to execute the *transition*. If the *request for quote* is refused the left *transition* leading to a *final state* is executed. In this case the collaboration failed on the business level. But if the *request for quote* succeeds - in other words if a *quote* is *provided* - we get to the next interaction called *place order*. This *business transaction activity* is again refined by a *business transaction* that deals with the submission of an *order*. If the *order* is successfully placed the corresponding *business entity* capturing the *order* status switches to state *accepted*. The *transition* that requires the *order* to be *accepted* is executed and we reach the *final state* called success. However, if the order placement fails in the respective transaction the *order business entity* gets in a state *rejected*. According to its flow, the collaboration consequently fails.

7.3.1.1 Artifacts

Modeling the *business choreography view* results in collaborative processes specified by **business collaboration protocols**. *Business collaboration protocols* contain a flow of *business transactions* and nested *business collaboration protocols* that describe the modeled process in detail.

7.3.2 Business Interaction View

7.3.2.1 Overview and purpose

The *business interaction view* captures the sequence of an interaction on the lowest level of granularity. Such interactions cover an information exchange between exactly two roles, consisting of a request from the initiating role and an optional response from the reacting role.

In the UMM interactions are defined by the abstract concept *business interaction behavior*. A *business interaction* is a persistent representation of a *business interaction behavior* that allows to capture the current state of its execution. In this version of the UMM, a *business transaction* is the only specialization of a *business interaction behavior* choreographing an information exchange. Other approaches of a *business interaction behavior* may be developed in a future version.

Business transactions describe an information exchange

The requirements on a *business transaction* are captured by a corresponding *business transaction use case* located in a *transaction requirements view* in the BRV. The *business transaction use case* defines the two participating roles and identifies the purpose of the *business transaction* as well as the actions performed.

Requirements are captured by business transaction use cases

A *business transaction* refines a *business transaction activity* that is part of a *business collaboration protocol*. In order to execute a *business transaction activity*, the sequence of the refining *business transaction* is started. The first step in a *business transaction* is a *requesting business activity* performed by the initiating role. The *requesting business activity* outputs an envelope that is transmitted to the corresponding partner. When the reacting role receives the envelope an activity is started. This *responding business activity* processes the received envelope. If required by the workflow, a response envelope is created by the *responding business activity* and sent to the initiating role. The initiating role receives the envelope and processes it. In case of a response the *business transaction* terminates when the initiating role receives and finishes processing the response envelope. Otherwise the *business transaction* is completed when the reacting role receives and finalizes processing the requesting envelope.

Choreography of a business transaction

The exchange of information performed by a *business transaction* is needed to synchronize the states of affected *business entities*. If a business entity state is of common interest, the partner that recognizes the state change initiates the *business transaction*. The execution of a *business transaction* leads to synchronized states in both partners' systems. If the initiating partner informs its counterpart about an irreversible state (e.g. a dunning letter, an order was shipped, etc.) a one-way transaction is performed. The corresponding partner has to accept the state change of a *business entity*. Otherwise, if an action of the other partner is required (e.g. price request, order submission, etc.) a two-way transaction is used. In case of a two-way transaction, the state of a *business entity* is set to an interim state when the requesting envelope is transmitted and set to a final state by the reacting partner. In this case the reacting partner decides about the final state. After a *business transaction* is completed the state of affected *business entities* is irreversible. If the previous states of *business entities* need to be reestablished, another compensating *business transaction* is needed. A simple rollback, as common in a database environment, is not possible in the interaction of e-business systems.

Synchronizing the states of business entities between business partners

Each one-way and two-way *business transaction* follows the corresponding workflow schema described above. But real-world business transactions differ in their purpose of information exchange (e.g. price query, order submission, etc.). This results in a different level of importance that in turn results in different requirements regarding security and legal aspects. In order to cover these requirements *tagged values* are used to parametrize security facets and legal aspects.

Tagged values specify different requirements on business transactions

UMM defines six, two one-way and four two-way, business transaction patterns that are suitable for every real-world transaction purpose (e.g. *commercial transaction, notification*, etc.). These patterns differentiate from each other in different *tagged value* settings resulting in different legal and security constraints. If required, a business transaction pattern can be further customized by altering certain *tagged value* settings. The six business transaction patterns are further detailed in *Appendix - Business Transaction Patterns*.

7.3.2.2   Stereotypes

BusinessInteraction (Class): A *business interaction* is a persistent representation of the execution of a *business interaction behavior*. In this version of the UMM, *business transactions* are the only valid concept to describe a *business interaction behavior*.

BusinessTransaction (ActivityGraph): A *business transaction* describes an information exchange between exactly two participants including a request from the initiating role and an optional response from the reacting role. *Business transactions* are used to synchronize the states of business entities between two partners. If the initiating partner informs its counterpart about an irreversible state (e.g. notification of shipment), a one-way transaction is used. Then the corresponding partner has to accept the state of a business entity. Otherwise, if an action of the other partner is required (e.g. price request, order submission, etc.) a two-way transaction is used. The state of a business entity is set to an interim state when the requesting envelope is transmitted. As soon as the reacting partner communicates its response, the business entity's final state is set. A *business transaction* is an atomic interaction. If a previous state of a business entity must be recovered, another compensating *business transaction* is needed.

BusinessTransactionSwimlane (Partition): A *business transaction swimlane* defines an area of responsibility. The contents of a *business transaction swimlane* are performed by a certain *authorized role*. The *authorized role* is assigned as a *classifier* to the *business transaction swimlane*.

RequestingBusinessActivity (ActionState): A *requesting business activity* is performed by the *authorized role* that initiates a *business transaction*. A *requesting business activity* outputs a *requesting information envelope* to the reacting role. In case of a two-way transaction the *requesting business activity* consumes the *responding information envelope* communicated by the reacting party.

RespondingBusinessActivity (ActionState): A *responding business activity* is performed by the *authorized role* that reacts in a *business transaction.* The *responding business activity* consumes the *requesting information envelope* sent by the initiating party. In a two-way transaction the *responding business activity* outputs a *responding information envelope* to the initiating party.

RequestingInformationEnvelope (ObjectFlowState): A *requesting information envelope* contains business information sent from the initiating role to the reacting role. The information contained in an information envelope leads to a change of one or more business entities. The term *requesting information envelope* does not mean that the business information refers to a request in a business sense. The term *requesting information envelope* indicates that the execution of a transaction is requested from the requesting role to the responding role - no matter whether this is an information distribution, a notification, a request, or the offer in a commercial transaction.

RespondingInformationEnvelope (ObjectFlowState): A *responding information envelope* is sent back from the reacting role to the initiating role in the case of a two-way transaction. The information contained in the *responding information envelope* sets one or more business entities from an interim state to their final state

7.3.2.3   Worksheets

Business interaction view worksheet

Common Information
Business library information

Business transaction worksheet

Common Information
Definition: Describes the overall customer value that is created by the business transaction for both participants.
Business transaction pattern: Specifies the business transaction pattern used in the particular *business transaction.* UN/CEFACT mandates the use of one of the following six patterns (*Appendix - Business Transaction Patterns*) that have been defined in the *RosettaNet framework*:

> Commercial transaction
> Request/Confirm
> Request/Response
> Query/Response
> Information Distribution

Notification

Secure transport: Indicates whether the message exchange has to be executed via a secured channel or not. The secure channel ensures that the content of a document is protected against unauthorized disclosure and modification. This security facet applies only to the time of transmission of a document. Once a document is transmitted, it has no longer an impact. A secure transport channel has to fulfill the following requirements:

Authenticate sending role identity

Authenticate receiving role identity

Verify content integrity

Maintain content confidentiality

Information that captures the requirements of the requesting business activity (note: the role name and the activity name are already gathered by the corresponding business transaction use case)

Time to respond: Specifies the time period that this transaction must be completed within. In other words, it defines the time frame by which the responding party has to return the request document. In case of a one-way transaction, this value must be null. Otherwise, time to respond should be specified as XML schema duration type.

Number of retries: Specifies how often the requesting role has to re-initiate the transaction if a time out exception occurs. Exceeding the *time to acknowledge receipt,* the *time to acknowledge processing* or the *time to respond* causes a time out exception. However, this *retry count* does not cover exceptions due to erroneous document content or to a bad message sequence. Moreover it is noteworthy, that the number of retries does not include the first, regular transmission attempt.

Time to acknowledge receipt: Specifies the time period within the responding role has to acknowledge receiving the requestor's document. If no *acknowledgement of receipt* is required this value must be set to null. Otherwise the proper time value should be specified as XML schema duration type.

Time to acknowledge processing: Specifies the time period within the responding role has to acknowledge processing of the requestor's document. The *acknowledgement of processing* is sent, when the document passes a validation against a set of business rules and is handed over to the application for processing. If no *acknowledgement of processing* is required this value must be set to null. Otherwise the proper time value should be specified as XML schema duration type.

Authorization required: Specifies whether the responding role must authorize itself or not. If so, the responding role must sign the business document and the requesting role must validate the response document's signature and approve the sender.

Non repudiation required: True if the responding party must not be able to repudiate its business action as well as the business documents sent.

Non repudiation of receipt required: True if the responding role has to sign the *acknowledgement of receipt*. This must only be considered if an *acknowledgement of receipt* is required. The non repudiation of a receipt ensures that the responding party is not able to deny sending the receipt.

Intelligible check required: True if the responding role has to check if the received document is readable before returning an *acknowledgement of receipt*. In other words, the receiver has to check if the document structure and its content has not been garbled during transmission.

Information that captures the requirements of the responding business activity (note: the role name and the activity name are already gathered by the corresponding business transaction use case)

Time to acknowledge receipt: Specifies the time period within the requesting role has to acknowledge receiving the responder's document. If no *acknowledgement of receipt* is required this value must be set null. Otherwise the proper time value should be specified as XML schema duration type.

Time to acknowledge processing: Specifies the time period within the requesting role has to acknowledge processing the responder's document. The *acknowledgement of processing* for a business document is sent, when the document passes a validation against a set of business rules and is handed over to the application for processing. If no *acknowledgement of processing* is required this value must be set null. Otherwise the proper time value should be specified as XML schema duration type.

Authorization required: True if the requesting role must authorize itself. If so, the requesting role must sign the business document and the responding role must validate the request document's signature and approve the sender.

Non repudiation required: True if the requesting party must not be able to repudiate its business action as well as the exchanged business documents.

Non repudiation of receipt required: True if the requesting role has to sign the *acknowledgement of receipt*. This information is only considered if an *acknowledgement of receipt* is required. The non repudiation of a receipt ensures that the requesting party is not able to deny sending the receipt.

Intelligible check required: True if the requesting role has to check if the received document is readable before returning an *acknowledgement of receipt*. In other words, the requesting role has to check if the document structure and its content has not been garbled during transmission.

The following information describes security requirements concerning the exchanged business documents. This information is relevant for each business document exchanged in the business transaction. Thus, the information below has to be gathered twice in case of a two-way transaction, but only once for a one-way transaction.

Information type: Denotes the business document's type.

Information state: Denotes the state a corresponding business entity is set to as a result of the business transaction's execution.

Are contents confidential: Defines if the message has to be encrypted in order to secure the information from unauthorized access.

Is the envelope tamper proof: Specifies if the sender must digitally sign the business document. A digital signature is a signed message digest that allows to check if the document has been tampered with.

Authentication required: Specifies if the sender's digitial certificate has to be associated with the transmitted document in order to proof it's identity.

7.3.2.4 Step by step modeling guide

1. Model the *business transaction* and its persistent representation
2. Determine the business transaction pattern
3. Model the *business transaction swimlanes* for both participants
4. Describe the requestor's part
5. Describe the responder's part

The choreography of a *business transaction* follows always the same structure. A minor difference results from the fact that the information exchange is two-way or just one-way. Thus describing a *business transaction* follows always the same modeling steps.

Tab. 7–10 Order from quote example: worksheet for the business transaction place order

Form: BusinessInteraction	
General	
Business Transaction Name	Place Order
Definition	The purpose of this business transaction is to provide a means for a buyer to send to the seller a list of items that they wish to purchase, and to provide a means for a seller to send to the buyer a formal acceptance or rejection of the buyer's order.
Select Business Transaction Pattern	CommercialTransaction
Secure Transport	true
Requestor's Side	
Requesting Role	Buyer
Requesting Business Activity Name	submit Order
Time to Respond	PT24H
Time to Acknowledge Receipt	PT2H
Time to Acknowledge Processing	PT6H
Authorization Required	true
Non Repudiation Required	true
Non Repudiation of Receipt Required	true
Intelligible Check Required	true
Number of Retries	3
Responder's Side	
Responding Role	Seller
Responding Business Activity Name	process Order
Time to Acknowledge Receipt	PT2H
Time to Acknowledge Processing	PT6H
Authorization Required	true
Non Repudiation Required	true
Non Repudiation of Receipt Required	true
Intelligible Check Required	true
Business Information Envelopes	
Information Envelope from Requesting Business Activity	
Information Name	PurchaseOrder
Information State	
Are Contents Confidential?	true
Is the Envelope Tamperproof?	true
Authentication Required?	true
Information Envelope from Responding Business Activity	
Information Name	PurchaseOrderResponse
Information State	
Are Contents Confidential?	true
Is the Envelope Tamperproof?	true
Authentication Required?	true

Start by adding a *business interaction view* to the *business transaction view*. One *business interaction view* describes exactly one *business transaction*, hence one *business interaction view* is needed for each *business transaction*.

Model the business transaction and its persistent representation

Similar as for *business collaboration protocol*s, the state of execution of a *business transaction* is captured by a persistent representation. The persistent representation of a *business transaction* is described by a *business interaction*. Hence, we add a *business interaction* to the *business interaction view*. Then model the *business transaction* as a child of the *business interac-*

A business interaction captures the state of a business transaction

tion denoting that the *business transaction* specifies the behavior of the *business interaction*.

We demonstrate modeling of *business transactions* by the example of the *place order business transaction* (of the *order from quote* example). Modeling all other *business transactions* follows the same procedure as the *place order* transaction. As shown in Figure 7–20, we add a *business interaction view* called *place order* to the *business transaction view*. Within this *business interaction view* we add a *business interaction* (BI), which acts as the persistent representation. The actual *place order business transaction* (BT) is then modeled as a child of the *business interaction* (BI). This parent-child relationship indicates that the *business interaction* (BI) always refers to the current state of the *business transaction* (BT).

Fig. 7–20 Relationship between a business interaction (BI) and a business transaction (BT)

Determine the business transaction pattern

In UMM a *business transaction* follows one of six business transaction patterns. These patterns cover every real-world business case which results in a legally binding interaction between two decision making applications as defined in Open-edi [OER95]). UN/CEFACT mandates to classify each UMM *business transaction* according to these patterns, which have proven to be useful in *RosettaNet* [ROS02] [Bir05]. Within these six business transaction patterns we differentiate between two one-way and four two-way patterns.

UMM provides six business transaction patterns

We first concentrate on one-way transactions. If the business information sent is formal non-repudiable it is called *notification*. Otherwise if just informal information is transmitted we are talking about an *information distribution* transaction.

One-way: information distribution and notification

In a two-way transaction the initiating side might request (static) information that the corresponding site has already available. In this case the transaction is called *query/response*. Otherwise, if response information is required that needs to be dynamically assembled or which is not available at the time of request, the transaction follows the *request/response* pattern. A *request/confirm* transaction differs from *request/response* in requiring a confirmation by the responder to the request of the initiator. Finally, a *commercial transaction* results in a residual obligation between both parties to fulfill the terms of a contract. This pattern represents the common „offer and acceptance" interaction.

Two-way: request/response, query/response, request/confirm, commercial transaction

The appropriate pattern has to be specified in the tagged value *business transaction type*. Further information to business transaction patterns is available in the annex (*Appendix - Business Transaction Patterns*)

Model the business transaction swimlanes for both participants

Each *business transaction* has exactly two *business transaction swimlanes*, one for the initiating role and one for the reacting role. These two *partitions* contain all *activities* and *exchanged envelopes* modeled in subsequent steps. Each *business transaction swimlane* must have a *classifier* assigned that corresponds to the role performing the *activities* located in the specific *partition*. The role must be one of two *authorized roles* participating in the corresponding *business transaction use case* (that captures the requirements of the *business transaction*). Thus each *authorized role* of the *business transaction use case* is assigned as a *classifier* to one of the two *business transaction swimlanes*. Of course, a *business transaction swimlane* has just one *classifier*. You should conform to the convention to model the *business transaction swimlanes* vertically, whereas the left *partition* belongs to the initiating role and the right *partition* describes the actions of the reacting role. You may also optionally assign a name to the *business transaction swimlanes* (e.g. initiator, reactor, etc.). This naming has no further implications, but facilitates its understanding.

Add a business transaction swimlane for each of the two roles

The worksheet shown in Table 7–10 provides the necessary input to construct the corresponding *place order business transaction* (Figure 7–21). Since a *business transaction* is always performed between two participants, we add two *business transaction swimlanes* to the *business transaction*. Considering the worksheet shown by Table 7–10, the *buyer* is the requesting role in the business transaction and the *seller* is the responder. Both roles (*buyer* and *seller*) are already defined as participants of the corresponding *business transaction use case* (see Figure 7–15). In order to denote that the *buyer* performs the requesting part of the transaction, we define the *authorized role buyer* as *classifier* of the left *business transaction swimlane*. Similarly we specify the *seller* - being the corresponding *authorized role* - as *classifier* of the right *business transaction swimlane*. This structure denotes that actions of the left *business transaction swimlane* are performed by the *buyer* and similarly actions of the right *business transaction swimlane* are conducted by the *seller*.

The place order transaction of the order from quote example

Describe the requestor's part

Now model the contents of the initiating role's *partition*. Place an *initial state* and a *requesting business activity* somewhere in the upper region of the *partition* and connect them via a *transition* leading from the *initial state*

Describe the action of the initiating role

to the *action state*. Then place a *requesting information envelope* in the lower region of the *business transaction swimlane*. A *requesting information envelope* represents an instance of an *information envelope*. Thus assign the *information envelope* that is sent in this step as a *classifier* to the *requesting information envelope*. This implicates that a message of the *information envelope* type is transmitted by the initiating role. Assigning a name to the *requesting information envelope* has no further relevance for the description, so you may skip it. Connect the *requesting business activity* and the *requesting information envelope* using a *transition*, leading from the *activity* to the *envelope*.

Considering the requestor's side of the *place order business transaction* we start with modeling the *initial state* followed by the *requesting business activity*. As captured by the corresponding worksheet (Table 7–10) we denote the *activity submit order*. The *initial state* is connected to the *submit order activity* via a *transition* leading from the *initial state* to the *activity*.

Modeling the buyer's part of the place order business transaction

Now we model the information that is submitted from the requestor (*buyer*) to the responder (*seller*) by placing a *requesting information envelope object flow state* in the *buyer's business transaction swimlane*. According to the information gathered by the *business transaction* worksheet, the *buyer* transmits a *purchase order envelope*. In order to indicate this, the *information envelope* called *purchase order envelope* (denoted after the colon in the *requesting information envelope*) has to be defined as the *classifier* of the *requesting information envelope*. However, modeling *information envelopes* is done in the *business information view*, hence there are no *information envelopes* described in this step. You might stay in the *business interaction view* and classify the objects in the swimlanes with the appropriate *information envelopes* after completing the *business information view*. The other alternative is switching for every required *information envelope* to the *business information view*.

The buyer submits a purchase order envelope

Furthermore we need to draw a *transition* from the *requesting business activity* (*submit order*) to the *requesting information envelope*. This means, that the *submit order activity* outputs a document which is of type *purchase order envelope*. This document is transmitted to the *seller*.

Finalizing the buyer's part of the place order transaction

Fig. 7–21 Order from quote example: business transaction place order

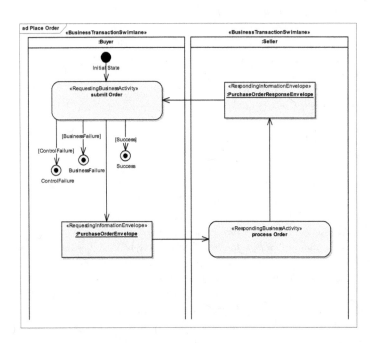

Describe the responder's part

Modeling the *business transaction swimlane* of the initiating role is completed so far, hence switch to the *business transaction swimlane* of the responding partition. Describe the reacting activity by placing a *responding business activity* in the lower region of the reacting role's *partition*. Add a *transition* leading from the *requesting information envelope* to the *responding business activity*. This constellation indicates that the *requesting business activity* sends a *requesting information envelope* to the *responding business activity*.

Model the reactor's action

Further modeling steps of a *business transaction* depend on the *business transaction*'s type. If the *business transaction* is a one-way transaction, add a *transition* leading from the *responding business activity* to the *requesting business activity*. Modeling this *transition* is not mandatory, it is recommended to facilitate understanding of the transaction flow.

One-way information exchange

If it is a two-way *business transaction* an envelope is transmitted back to the initiator. Therefore add a *responding information envelope* to the upper region of the reactor's *partition*. Followed by a *transition* leading from the *responding business activity* to the *responding information envelope* and one from the *responding information envelope* to the *requesting*

Two-way information exchange

business activity. The modeled flow indicates the transmission of a response from the reacting role to the initiating role.

Regardless of one-way or two-way transactions a *business transaction* must have at least two *final states.* Each *final state* must be located in the initiator's *partition* and be the target of a *transition* starting from the *requesting business activity.*

Add at least two final states

In our *place order* example we have a two-way transaction because the responder (*seller*) answers the order submission with a *purchase order response envelope.* Furthermore, our example transaction follows the *commercial transaction* pattern. It covers the typical offer/acceptance process that results in an residual obligation between two parties fulfilling the terms of a contract.

The place order example follows the commercial transaction pattern

In our *place order* example (Figure 7–21) we finally construct the responder's part which is carried out by the selling role. Regarding the worksheet (Table 7–10) the *seller* executes an activity called *process order.* Hence, we add a *responding business activity* called *process order* to the *seller's business transaction swimlane.* Then we connect the *requesting information envelope* sent by the requestor (*buyer*) with the *seller's process order activity.* This indicates that the *requesting information envelope* is input to the *seller's process order activity.*

Describing the responding business activity of the place order transaction

Since it is a two-way transaction the *seller's process order activity* also outputs a *responding information envelope* that is transmitted back to the buying role. According to the worksheet information the response document is of type *purchase order response envelope.* This requires that we add the *purchase order response envelope* (which is itself an *information envelope* and in the example figure denoted after the colon in the *response information envelope*) as *classifier* to the *response information envelope* object. Similarly to the requestor's side, *information envelopes* are not defined yet. Thus, this might be accomplished after finishing the *business information view* or switching between defining transactions and modeling exchanged information is required.

The seller returns an purchase order response envelope

In order to finish our example, we connect the *process order activity* with the *purchase order response envelope* via a *transition* leading from the *activity* to the *envelope.* Then we add a *transition* from the *purchase order response envelope* to the *buyer's submit order activity.* Finally, we add three *final states* to the requestor's side - one for the case of a *success,* one for a *control failure* and one for a *business failure.*

Finalizing the place order example

7.3.2.1 Artifacts

The *business interaction view* delivers a detailed description of interactions between exactly two *authorized roles.* Interactions are specified by the concept of **business transactions**.

7.3.3   Business Information View

7.3.3.1   Overview and purpose

The *business information view* covers the information that is exchanged in *business transactions*. In a *business transaction*, instances of *information envelopes* are transmitted between business partners that are either specified as *requesting* or *responding information envelopes*. An *information envelope* encapsulates information that is needed to synchronize the states of business entities. The UMM encourages to exchange only the minimal information needed to change business entity states. UMM also supports traditional document-centric approaches, but it is not recommended to use them.

The business information view describes the information exchanged within business transactions

    There are no restrictions on methods and rules for modeling the contents of an *information envelope*, but UN/CEFACT recommends to use the UMM specialization module for modeling Core Components (CCTS). This specialization module is currently under development and describes how Core Components have to be modeled in UML. Chapter 10 of our thesis discusses information modeling in UMM and describes this upcoming specialization module in more detail.

7.3.3.2   Stereotypes

InformationEntity (Class): An *information entity* describes a semantical unit of information that is transmitted between two partners in a *business transaction*. *Information entities* may be associated with other *information entities* describing more complex information structures.

InformationEnvelope (Class): An *information envelope* is a container for *information entities* that are exchanged in a *business transaction*. *Information envelopes* are composed only of *information entities*. Exactly one *information entity* takes on the role *header* and one or more *information entities* take on the role *body*. The information exchanged in a *business transaction* is always of type *information envelope*.

7.3.3.3   Step by step modeling guide

1. Describe the information envelope
2. Identify and specify required header information
3. Describe the business information

In this last step of the modeling workflow, specify the structure of messages that are exchanged in an interaction between business partners. Artifacts that describe a message are modeled in a *business information view*. UMM makes no restrictions if one or more *business information views* are used to structure the information modeling. However, a *business information view* must not contain any other model elements than *information envelopes* and *information entities*, but you may use *class diagrams* to ease the information modeling.

Going through the *business information view* and describing the information that is exchanged in *business transactions* is not guided by worksheets. This results from the fact that worksheets capture only the type of information that is needed and the relationships between pieces of information. However, in this respect worksheets provide no additional or supportive value. We think relevant information should be described directly in the *business information view* without using worksheets. Information modeling and supportive means therefor are relevant to further research (See "Mapping Business Information to Document Formats" on page 228).

The business information view is not supported by worksheets

Modeling a business document starts with defining an *information envelope*. An *information envelope* is a container for the actual business information and corresponds to the envelope of the business message. Security requirements on *information envelopes* may be specified by its properties *is authenticated*, *is confidential* and *is tamper proof*. The business information contained in an *information envelope* is described by *information entities*. Thus an *information envelope* is composed of *information entities* that may be also structured recursively.

An information envelope consists of one header and one to many body parts

An *information envelope* requires to have some header information associated within. This header information does not relate to the technical transmission level, but to business semantics as defined by UN/CEFACT'S Standard Business Document Header [SBD04]. The header of an *information envelope* is specified by exactly one *information entity*. Thus add one *information entity* and define the required header information by its properties. For each property define its name and its type. After describing the header *information entity*, connect it with the *information envelope* via an *aggregation* association. Finally, assign the role *header* to the *information entity* in order to define it as the envelope header.

As next step the actual business information is modeled using *information entities*. An *information entity* corresponds to a semantical unit of information and may define a set of properties to describe the business information. Model *information entities* recursively in order to describe more complex information structures. Finally, connect top-level *information entities* with their enclosing *information envelope* via an *aggregation* association. Assign the role name *body* to these *information entities* in order to define them as the actual business information.

The actual business information is denoted as the body of an information envelope

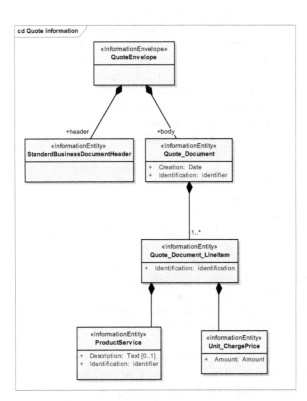

Fig. 7–22 Order from quote example: structure of the quote envelope

Figure 7–22 shows the structure of the *quote envelope*, which is exchanged in the *request for quote* transaction of our *order from quote* example. UMM requires that an *information envelope* is associated with exactly one *information entity* defined as *header* and with at least one *information entity* defined as *body*. In our *quote envelope* example, the header block is specified by an *information entity* named *standard business document header*. It is aligned to UN/CEFACT's *Standard Business Document Header*, which is not described in detail in this thesis. The body of the *quote envelope* is specified by an *information entity* named *quote document*. It represents the actual business document. A *quote document* contains one to many *quoted line items*. Each *line item* is uniquely identified by its *identification* attribute. Furthermore a *line item* has the actual *product service* and its *unit price* associated. A *product service* includes an *identifier* and an optional free text *description*. The *unit charge price* includes an *amount* attribute.

7.3.3.1 Artifacts

The *business information view* describes the information that is exchanged in *business transactions*. *Class descriptions* are utilized, whereby ***information envelopes*** are used as containers for the actual business information.

8   UMM Validator

One of the most important issues an UMM modeler is facing, is the chance   *Model validity is a key issue*
to determine if a model is valid or not. Although the UMM meta model,
which is based on a UML Profile, and the UMM business transaction patterns are helping the modeler to create a valid model, a final overall validation of the model is required.

A model which has been validated by a validator guarantees that applications which are using the model, are operating properly. Figure 8–1 shows
the UMM validator and its context within the model creation workflow.
With the help of the UML Profile for UMM the user creates an UMM
model, following the guidelines of the UMM meta model. After having finished the model, the user validates it with the help of the validator. If the validator does not return any error messages, the model can be used by further
applications such as a BPSS or a BPEL transformer.

*Fig. 8–1 The validator
and its context*

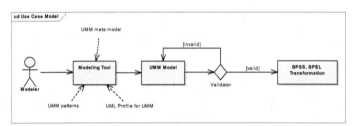

The following paragraphs will explain the motivation and the details as well
as the problems which occurred during the realization of the UMM validator.

8.1   Motivation for UMM validation

The diffusion of UMM knowledge is not as high as the diffusion of UML
knowledge. Often even a keen UML modeler is facing problems when creating an UMM compliant model. Equipped with the UMM meta model, the
UML Profile for UMM and the UMM patterns as well as the UMM specifi-

cation, the modeler creates an UMM compliant model. Whether the model is correct or not depends on the experience of the modeler. Nevertheless even the most experienced modeler can overlook a fault in the model. Exactly at this point the chance for validating the created model would bring relief to a modeler who is not entirely familiar with UMM.

However the validator itself does not only target at beginners and people not that familiar with UMM. Even UMM professionals sometimes oversee interdependencies between modeling elements during the validation process. Especially connectors between model elements are often invalid in a given model, but because they are invisible, the modeler does not perceive the apparent, but invisible error. In particular in Enterprise Architect the deletion of connectors can set them just on invisible instead of removing them. Although not particularly crucial in a model which is just intended to be a graphical representation perceivable for humans, invisible connectors are problematic, if the model is used by automatic processing such as a BPSS transformer.

The main aim of the UMM Validator can be divided into two significant points.

First the modeler should be guided towards a valid model. This means, that for instance after finishing a *business area* package within the *business domain view*, there should be the alternative to check, if the created *business area* is valid. Seen from the point of software/design engineering this is known as a bottom up approach. By gradually validating the different sub-packages, the modeler finally has a valid model. *Top down vs. bottom up validation*

The second main issue of the UMM validator is the overall validation of a given model. After having finished a model the user should have the chance to check the overall validity of the model. This approach is known as top down approach.

Both validation approaches are implemented by the validator. By applying the top down and the bottom up approach the modeler finally has a valid model, which is ready for further use. Moreover the creation of further applications which are using the created UMM model is extensively facilitated because a valid model can be anticipated.

Before we proceed to the details of the validator a short overview on *extension mechanisms* in UML will be given. UMM itself thoroughly uses the extensions mechanisms provided by the UML meta model.

8.2   UML Extensions

The Unified Modeling Language (UML) has certain extension mechanisms, which are widely used by UMM. Via *stereotypes* the different packages and elements are uniquely identified and *tagged values* help refining object properties. Finally *constraints* defined in the Object Constraint Language (OCL) restrict the users ability to model within a certain package. The following three paragraphs will give a brief survey of the three extension mechanisms mentioned above.

8.2.1 Stereotypes

The first extension mechanism to be examined are stereotypes. A stereotype represents a subclass of an element of the existing meta model. The stereotype has the same form, relationship and attributes as the existing meta model element, but the intent of the stereotype is a different one. By using a stereotype, one can indicate a distinction of usage within the model. Additionally the stereotype can have further constraints and tagged values in order to add information. A deeper insight into the concept of stereotypes and their usage is given by the UML specification [UMa04]. *Stereotypes are fundamental for UMM*

Within the UMM stereotypes play a very important role. Via stereotypes the different packages like the *business domain view* or the *business transaction view* are differentiated in the model. Without the help of stereotypes, it would not be possible for a modeler to indicate unambiguously, which package refers to which view.

To indicate, which element or package uses a stereotype, the modeler places the name of the stereotype surrounded by guillemets before the name of the element.

«BusinessTransactionView» OrderFromQuote *Example package structure*

    «BusinessInteractionView» PlaceOrder
    «BusinessInformationView» OrderInformation

The example above shows a *business transaction view* named *order from quote* with two subpackages named *place order* and *order information* and their dedicated stereotypes *business interaction view* and *business information view*.

If neither the package *place order* nor the package *order information* would have an associated stereotype, it would not be possible to evaluate, which package is the *business interaction view* and which package is the *business information view*.

As we will see in the following chapters, the validator thoroughly uses stereotypes to determine which view is meant and which subroutine within the validator should be called. The second extension mechanism also extensively used within the UMM are tagged values.

8.2.2 Tagged values

Tagged values can be compared to regular attributes as used in class elements. However attributes are defined in the M1 layer whereas tagged values are defined in the M2 layer of the Meta Object Facility (MOF). A deeper insight into the Meta Object Facility is given by [MO102]. The attributes in a class are giving a certain meaning to the objects derived from the class. They add information to the class and are visible to the user.

However often the modeler wants to add information via using attributes to other UML elements like packages and activity partitions too. These elements do not have attributes like the class elements have and therefore ancillary information would normally not be possible. Yet due to the tagged value mechanism users can add information to any element.

Although similar to attributes in class elements, tagged values are not the same. A tagged value is represented by a keyword-value pair which can be attached to an arbitrary element within the model - because it is defined in the M2 layer. In UMM the usage of tagged values is limited because the used stereotypes restrict the tagged values which may be used.

Figure 8–2 on the left hand side shows the definition of the two stereotypes *stakeholder* and *business partner* and their corresponding tagged values. In *stakeholder* one tagged value *interest* is defined. Because *business partner* is a subclass of *stakeholder*, it inherits the tagged values of the superclass. Furthermore (and not shown in this example) the subclass could define additional tagged values. The element *stakeholder* is of type *actor*. By definition it would normally not be possible to add additive information to the *stakeholder*. However with the tagged value mechanism it is possible to add additional information to the element. On the right hand side an actor *customs* is shown. It has the stereotype *stakeholder* and therefore the tagged value *interest* as well.

Fig. 8–2 Tagged value definition

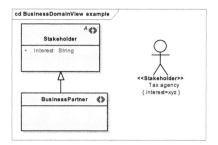

Because tagged values are central to the UMM, a special sub-validator has been developed, which exclusively validates the tagged values. We will have a deeper insight into that issue in chapter 8.5.4.

8.2.3   Constraints

As third and final extension mechanism of UML we discuss constraints and their use within the UMM. The UML meta model is related to a formal language called Object Constraint Language (OCL) which is used to express constraints. Constraints specify rules, which are called invariants. These invariants hold for a given model. If no rule is violated, the model is valid.

Motivation for OCL

Often a UML model element, such as an *action state* is not refined enough to meet all the requirements of a given specification. In this chapter the UMM specification is taken as an example.

Within the UMM specification additional requirements, which are not covered by the meta model are denoted in natural language. These requirements define additional constraints about the objects in the model. So far the additional constraints in natural language are a valid approach, as every modeler can understand them and apply them while modeling. Nevertheless practice has shown that natural language constraints often fail as they lead to ambiguities. With the use of a technical and unambiguous language this problem can be avoided. This is the point, where the OCL comes in. By capturing the constraints in OCL every modeler has the same and unambiguous base of understanding about the UMM.

Fig. 8–3 Sample from a collaboration requirements view

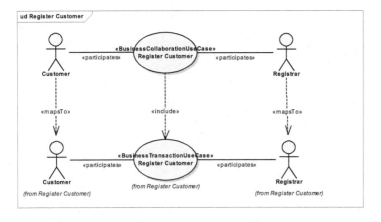

Figure 8–3 shows an example from a *collaboration requirements view*. One restriction on this view, expressed in natural language is the following:

> A source business collaboration use case includes target business transaction use cases and/or business collaboration use cases. Each role of the source use case must be mapped maximal once to a role of the same target use case (but it may be mapped to different authorized roles of different target use cases). Each role of the target use case is the supplier of a mapsTo dependency from a role of the source use case.

Natural language constraint for collaboration requirements views

It is apparent, that the restriction is quite complicated and not unambiguous. Therefore an OCL constraint is defined, which represents the constraint mentioned in natural language above.

```
[239] package Behavioral_Elements::Use_Cases
[240] context UseCase
[241]
[242] inv AuthorizedRoleofBTUCisSupplierOfOnlyOneAuthorizedRoleOfBCUC:
[243]    (self.isBusinessTransactionUseCase() or
[244]    self.isBusinessCollaborationUseCase()) implies
[245]    self.include->select(a | a.base <> self)->collect(base)->
[246]    collect(x | x.associations)->collect(y | y.allConnections)->
[247]    select(isAuthorizedRole)->forAll(x | self.associations->
[248]    collect(allConnections)->select(isAuthorizedRole)->
[249]    collect(supplierDependency)->collect(client)->isUnique(x))
```

Listing 8–1 OCL constraint for collaboration requirements views

Line 242 indicates the name of the so called invariant. The different invariants hold for the system which is being modeled. The sub-methods like *isBusinessTransactionUseCase()* are defined globally and are not quoted in the example above. By defining the constraints with OCL, the modeler is given a formal and unambiguous way for model restriction. For the time being only very few modeling tools support the direct validation of a model using constraints defined in OCL. Hence for the development of the UMM Validator a different though similar approach has been chosen.

Chapter 8.4 will give a deeper insight into UMM and OCL. Meanwhile the next chapter will focus on the UMM meta model, which is the base for the UMM validator.

8.3   The conceptual UMM meta model

A brief overview over UMM was already given in chapter 5. We will now focus on the conceptual UMM meta model, which is the base for the UMM validator. This chapter is based on the UMM meta model foundation module, published by UN/CEFACT [FOU03].

Within the conceptual model, the interdependencies between the different stereotypes and modules are defined. Figure 8–4 shows the composition of a *business collaboration model*.

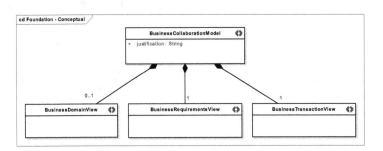

Fig. 8–4 A conceptual overview

A project which is UMM compliant is stereotyped as *business collaboration model*. A *business collaboration model* consists of zero or one *business domain view*, one *business requirements view* and one *business transaction view*.

The first view to be analyzed in detail will be the *business domain view*.

8.3.1   Business Domain View

The *business domain view* is used to discover and identify the processes, which are relevant for the *business collaboration model*. Figure 8–5 shows the *business domain view* at a glance.

Discover and identify processes

As we can see, a *business partner* participates in zero or more *business processes*. A *business process* itself is performed by one or more *business partners*.

Fig. 8–5 Composition of the business domain view

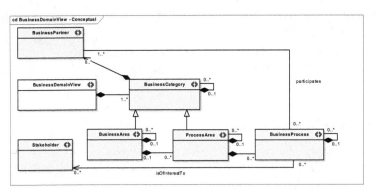

A *business partner* is a specialization of a *stakeholder* (not shown in this diagram). A *stakeholder* has an interest in multiple *business processes* and *business processes* might be of interest to multiple *stakeholders*. In a UMM compliant model this fact is denoted by an *is of interest to* dependency between the *business process* and the *stakeholder*. A *stakeholder* must not necessarily participate in the *business process*.

A *business process* can consist of sub-processes, which are associated with the super-process via *include* and *extends* associations.

In order to facilitate the identification of *business processes*, the processes are categorized using *business categories*. As denoted in Figure 8–5, a *business domain view* is composed of one or more *business categories*. A *business category* itself can be composed of other *business categories*, which allows the user to build a hierarchy.

A *business process* is assigned to exactly one *business category*. A *business category* on the lowest level of a *business category* hierarchy includes one or more *business processes* whereas a *business category* on a higher level does not include any *business process*.

The specializations of a *business category* are *business area* and *business process*. UN/CEFACT suggests their use, although it is not mandatory. A division within an organization corresponds to a *business area*. Common operations within a *business area* are aggregated in a *process area*. Like the *business category*, *business areas* and *process areas* can form a hierarchy.

Business areas may include only *business areas*. The only exception is the *business area* on the lowest level of the hierarchy, which is composed of one ore more *process areas*. *Business areas* must not include *business processes*.

The stereotype *business category* and the combination of the stereotype *business area* and *process area* are considered as alternatives. A UMM compliant model must not use both alternatives.

Subsequent to the *business domain view* is the *business requirements view*, which will be the topic of the next chapter.

8.3.2   Business Requirements View

The goal of the *business requirements view* is the identification of collaborative business processes between different business partners. Furthermore it describes the requirements of the identified business processes. As shown in Figure 8–6, the *business requirements view* consists of three different artifacts, which help to evaluate the requirements of a collaborative business processes. Namely the three artifacts are the *business process view*, the *business entity view* and the *partnership requirements view*. This chapter will give a short overview about the *business requirements view* as a whole.

Identification of collaborative business processes

The following chapters will then focus on the three mentioned artifacts in detail.

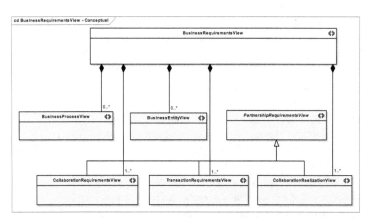

Fig. 8–6 Composition of the business requirements view

The flow of activities and states of *business processes*, which have been discovered in the *business domain view* is described in the *business process view*. The view itself is optional. Nevertheless there can be more than one *business process view* within a *business requirements view*.

The life cycles of *business entities* that are manipulated in a collaborative *business process* are described in the so called *business entity view*. Like the *business process view*, the *business entity view* is not mandatory and can occur more than once within a *business requirements view*.

The requirements on a partnership between *business partners* are covered by the *partnership requirements view*. On the lowest level of granularity, a partnership is a *business transaction*. *Business collaborations* are partnerships, which consist of *business transactions* and/or other *business collaborations*.

Requirements concerning a *business transaction* are covered by the *transaction requirements view*, those requirements concerning a *business collaboration* are covered by the *collaboration requirements view*.

The realization of a *business collaboration* can be executed between multiple sets of different *business partners*. The requirements concerning a realization of a *business collaboration use case* are covered by the *collaboration realization view*. A *collaboration realization* is specific for a set of *business partners*. As shown in the preceding illustration, the *partnership requirements view* is an abstract concept, which is either realized by the *collaboration requirements view*, the *transaction requirements view* or the *collaboration realization view*.

Within a model at least one *business collaboration,* containing a *business transaction* should be described. One of the *business collaborations* must then be executed by a set of *business partners.*

The *business requirements view* must contain at least one *collaboration requirements,* one *transaction requirements* and one *collaboration realization view.* The three mentioned views can occur multiple times within a *business requirements view.*

The first view to be analyzed in detail will be the *business process view.*

Business Process View

Figure 8–7 gives an overview about the *business process view* and the participating stereotypes.

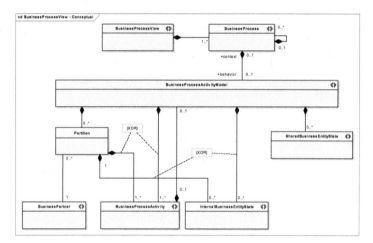

Fig. 8–7 The business process view at a glance

The *business process view* gives an overview about the *business processes,* the activities the processes consist of and the participating *business partners,* which execute the activities. The *business process view* consists of one or more *business processes.* In case that more than one *business process* is included, the *business processes* should be related. *Business processes* might include or extend other *business processes.* This fact is denoted by the unary composition assigned to the *business process.*

The dynamic behavior of *business processes* is described by the *business process activity model.* A *business process* is composed of zero or one *business process activity models.* Whether the flow of a *business process* is described by a *business process activity model* or not depends on the relevance of the *business process.* The *business process activity model* describes a flow of activities, which are performed by one or more participants. In case two or more *business partners* collaborate, the *business pro-*

Overview about business processes, activities and business partners cess activity model is divided into *partitions* where each *partition* is for one *business partner*. In case the *business process* is an internal business process, which is executed by one *business partner*, the *partition* for the partner is optional. Thus the *business process activity model* is composed of zero or more *partitions*. A *partition* is a UML standard element.

A *partition* is assigned to one *business partner* and a *business partner* is assigned to one *partition* in one *business process activity model*. Though, a *business partner* can be assigned to multiple *partitions* with the restriction, that each *partition* is in a different *business process activity model*.

Partitions and their corresponding business partners

A *business process activity model* is denoted by a flow of *business process activities*. In case no *partition* is used, the *business process activities* are directly included in the *business process activity model*. If *partitions* are used, the *business process activity* is assigned to the *partition* of the *business partner* who is executing the *business process activity*. In 8–7 this is shown by the *XOR* constraint on the left hand side. Whenever a transition connecting two *business processes activities* crosses the border between *partitions*, a collaborative business process is found. A *business process activity model* is composed of one or more *business process activities*, or a *partition* is composed of one or more *business process activities*. A *business process activity* can be refined by another *business process activity model*. Therefore a *business process activity* is composed of zero or one *business process activity models*, which then are a composite of zero or one *business process activities*.

Important states of *business entities* may also be described by a *business process activity model*. These states are manipulated during the execution of a *business process*. One can regard a *business entity state* as the output from one *business process activity* and the input to another *business process activity*. A transition from a *business process activity* to a *business entity state* indicates an output. Similarly a transition from a *business entity state* to a *business process activity* indicates an input. *Business entity states* which are meaningful to one *business partner* only are the so called *internal business entity states*. *Business entity states*, which must be communicated to a *business partner* are so called *shared business entity states*. Both kind of business entity states may be included in a *business process activity model*. Therefore a *business process activity model* may be composed of zero to many *internal business entity states* and *shared business entity states*. If *partitions* are used, the two *business process activities* which are creating and consuming an *internal business entity state* are in the same *partition*. Contrary the two *business process activities* which are creating and consuming a *shared business entity state* are in different *partitions*. A *shared business entity state* indicates the need for a collaborative *business process*.

Describing important business entity states

The next view focuses on the life cycles of *business entities* which are manipulated in a collaborative *business process*. It is the so called *business entity view*.

Business Entity View

Figure 8–8 gives an overview about the *business entity view* and its participating stereotypes.

Fig. 8–8 The business entity view at a glance

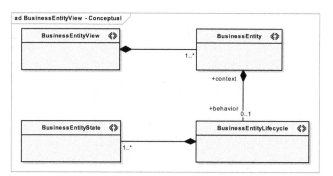

A real world representation which has business significance and is shared among two or more *business partners* in a collaborative *business process* is called a *business entity*. A collaborative *business process* can for instance be *purchase product*, *register customer* etc. Within a *business requirements view* multiple *business entity views* can occur, however there must be at least one.

Defining a business entity

The *business entity view* itself is composed of one to many *business entities*. Whether a *business entity lifecycle* is included in the *business entity view* or not depends on the importance of the *business entity lifecycle*. A *business entity* is therefore composed of zero to one *business entity lifecycles*. The different *business entity states* a *business entity* can have are described by a *business entity lifecycle*. Within a *business entity lifecycle* there must be at least one *business entity state*. Therefore a *business entity lifecycle* is composed of one or more *business entity states*. Because a *business entity lifecycle* is a UML *state machine* it can include events and transitions together with optional guards, which lead from one *business entity state* to another.

After having examined the constraints which apply to *business entities* we proceed to the *partnership requirements view* which handles the requirements on a partnership between *business partners*.

Partnership requirements view

Figure 8–9 gives an overview of the stereotypes in a *partnership require-ments view* and their interdependencies.

Fig. 8–9 Overview of the partnership requirements view

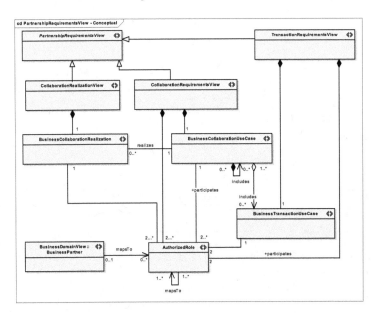

The *business entity view* and the *business process view* which were explained in the last two chapters help to identify the need for a collobora-tion.

The *partnership requirements view* denotes the requirements of an identified collaboration between *business partners*. These requirements are described by *use cases*. We distinguish between three stereotypes of use cases: *business transaction use case*, *business collaboration use case* and *business collaboration realization*. A *business transaction use case* describes the requirements of a *business transaction*. A *business transac-tion* is a special interaction between two roles. The interaction focuses on the initiating information exchange and an optional response. The require-ments of a *business collaboration* that is executed between two or more roles is described by a *business collaboration use case*. A *business collabo-ration* is composed of one or more *business transactions* or nested *business collaborations*. A set of *business partners* is required for the execution of a *business collaboration use case*. A *business collaboration use case* may be realized by different sets of *business partners*. The realization of a *business*

Requirements of a collabo-ration between business partners collaboration by a specific set of *business partners* is described by a *business collaboration realization*.

As Figure 8–9 shows, the *partnership requirements view* is an abstract concept, which can either be a *collaboration requirements view*, a *transaction requirements view* or a *collaboration realization view*. A *collaboration requirements view* captures the requirements of a *business collaboration* and a *transaction requirements view* captures the requirements of a *business transaction*. Finally a *collaboration realization view* captures the requirements of a *business collaboration realization*.

In each *collaboration requirements view* exactly one *business collaboration use case* is defined. The participants of a *business collaboration use case* are two or more *authorized roles*. In the *collaboration requirements view* package where the *business collaboration use case* is defined, the two corresponding *authorized roles* must be defined as well. Therefore a *collaboration requirements view* is composed of two or more *authorized roles*. If an *authorized role* participates in multiple *business collaboration use cases*, different *authorized roles* must be defined. That means, that each *authorized role* of the same role is in a different namespace of a corresponding *collaboration requirements view*. Accordingly an *authorized role* participates in only one *business collaboration use case* - it is the one in the same *collaboration requirements view*. The relation between the *business collaboration use case* and its *authorized roles* is 1 to (2..n), as shown in the figure above. No *authorized role* must be associated more than once to the same *business collaboration use case*.

Define business collaboration use cases in the collaboration requirements view

The *transaction requirements view* describes the requirements of a *business transaction*. In each *transaction requirements view* exactly one *business transaction* is described by a *business transaction use case*. Exactly two *authorized roles* participate in the *business transaction use case*. The *authorized roles* must be defined in the same package as the *transaction requirements use case* in which they participate in. Therefore a *transaction requirements view* is composed of exactly two *authorized roles*. That means, that each *authorized role* of the same role is in a different namespace of a corresponding *transaction requirements view*. Accordingly an *authorized role* participates in exactly one *business transaction use case* - it is the one in the same *transaction requirements view*. The relation between the *business transaction use case* and *authorized roles* is 1 to 2 as shown in the figure above. No *authorized role* is associated twice to the same *business transaction use case*.

Define business transaction use cases in the transaction requirements view

A *business collaboration use case* may include nested *business collaboration use cases*. Optionally a *business collaboration use case* can be nested in multiple parent *business collaboration use cases*. This fact is represented by the unary (0..n) to (0..n) include composition in the figure

Use cases can be nested

above. Multiple *business transaction use cases* may be included in a *business collaboration use case*. A *business transaction use case* must be included in at least *one business collaboration use case*. In the figure above this can be seen in the (1..n) to (0..n) aggregation between the *business collaboration use case* and the *business transaction use case*. A *business collaboration use case* includes at least one use case - no matter whether the use case is a nested *business collaboration use case* or a *business transaction use case*. No cycles must be included by a hierarchy of *business collaboration use cases* built by include compositions. A *business transaction use case* cannot be further disassembled by an *include* association. No *extend* associations between business collaboration/transaction use cases are used in UMM.

For every *include* relationship between a *business collaboration use case* and a *business transaction use case* as well as for a relationship between two *business collaboration use cases*, a mapping of the *authorized role* of the source use case to the *authorized roles* of the target use case is necessary. Therefore the *authorized role* has a (1..n) to (1..n) *mapsTo* relationship. Every role of the source use case may be mapped at maximum once to a role of the same target use case. Nevertheless a role may be mapped to different *authorized roles* of different target use cases.

Mapping of authorized roles

The *business partners* which have been identified in the previous UMM steps must not directly be associated with the *business collaboration use cases* and the *business transaction use cases*.

In order to indicate, that a specific set of *business partners* collaborate, the concept of a *business collaboration realization* is used. Every *business collaboration realization view* defines exactly one *business collaboration realization*. A *business collaboration realization* realizes exactly one *business collaboration use case*. A *business collaboration use case* can be realized by multiple *business collaboration realizations*. It is not necessary, that each *business collaboration use case* has a corresponding *business collaboration realization*. Therefore the *realize* association between a *business collaboration use case* and a *business collaboration realization* is 1 to (0..n).

Define specific collaboration sets in the collaboration realization view

Two or more *authorized roles* participate in a *business collaboration realization*. It is necessary, that these *authorized roles* are defined in the same package as the *business collaboration realization*. Therefore a *collaboration realization view* is composed of two or more *authorized roles*. Normally the names of the *authorized roles* which participate in the *business collaboration use case* are the same as the names of the *authorized roles* in the *business collaboration realization*, realizing it. Nevertheless the *authorized roles*, which participate in the *business collaboration use case* and in the *business collaboration realization* will be defined in different namespaces - each in the corresponding view. The *business collaboration realization* and the *authorized roles* are related by an 1 to (2..n) association. The number of actors which participate in the *business collaboration realization* must be the same as the number of actors participating in the *business collaboration use case*, which is realized by the *business collaboration realization*.

A *business collaboration realization* is associated with the *business partners* executing it by mapping the *business partners* to the *authorized roles* participating in the *business collaboration realization*. Each *authorized role* of a *business collaboration realization* is target of exactly one *mapsTo* association from a *business partner*. A *business partner* may only map to one *authorized role* in a *business collaboration realization*, but it may map to several *authorized roles*, as long as they are in different *business collaboration realizations*. In the figure above this is shown by the (0..1) to (0..n) *mapsTo* association between *business partner* and *authorized role*.

With this view the conceptual description of the *business requirements view* is finished, and we continue with the *business transaction view*.

8.3.3 Business Transaction View

The *business transaction view* shows how the business analyst sees the process to be modeled. It is an elaboration of the *business requirements view* by the business analyst. The *business transaction view* shows the choreography of information exchanges according to the requirements of the *business requirements view*. Within the *business transaction view* there are three artifacts, which together describe the overall choreography of information exchanges. As shown in Figure 8–10 these three artifacts are *business choreography view*, *business interaction view* and *business information view*.

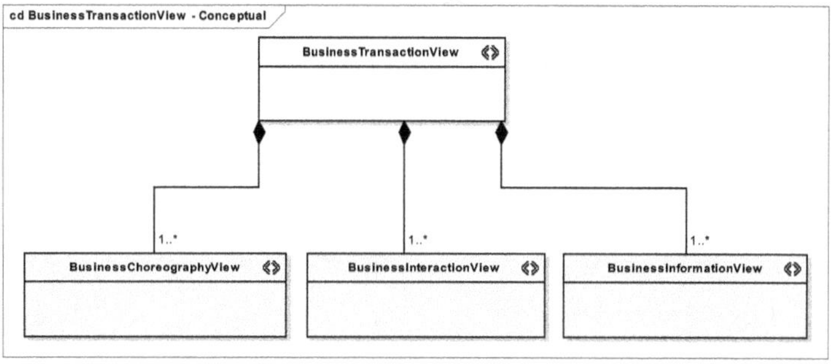

The business analysts view on the process

Fig. 8–10 The business transaction view at a glance

The *business choreography view* contains artifacts, which describe the flow of complex *business collaborations* between *business partners*, that may involve many steps. In detail the *business choreography view* contains artifacts which define a flow according to the requirements which have been evaluated in the corresponding *collaboration requirements view* of the *business requirements view*.

The *business interaction view* contains artifacts, which define a choreography which in turn leads to synchronized states of *business entities* on both sides of the interaction. In detail the *business interaction view* contains artifacts, which capture a flow according to the requirements of the corresponding *transaction requirements view* of the *business requirements view*.

The *business information view* contains artifacts, which describe the information exchanged in an interaction. Dynamic aspects of a collaboration are described by artifacts in the *business choreography view* and in the *business interaction view*. The *business information view* contains artifacts which describe the structural aspects of a collaboration. A *business transaction view* must contain at least one *business choreography view*, one *business interaction view* and one *business information view*. However, the three mentioned views can also occur multiple times.

The first subview of the *business transaction view* to be examined in detail is the *business choreography view*.

Business Choreography View

The main purpose of this view is to define the *business choreography* of exactly one *business collaboration*. Therefore the *business choreography view* is composed of exactly one *business choreography*, which is a persistent representation of the execution of a *business collaboration*. As shown in Figure 8–11 the execution order of a *business collaboration* is defined by the *business choreography behavior*.

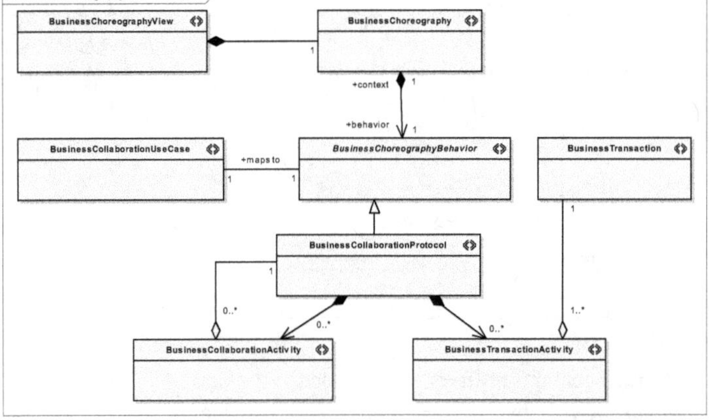

Define the choreography of a business collaboration

Fig. 8–11 Overview of the business choreography view

A *business choreography* is composed of exactly one *business choreography behavior*. The *business choreography behavior* follows exactly the requirements, which are defined in the corresponding *business collaboration use case* of the *business requirements view*. Furthermore each *business collaboration use case* of the *business requirements view* is mapped to exactly one *business choreography behavior*. In the figure above this is shown by the 1 to 1 *mapsTo* relationship between the *business collaboration use case* and the *business collaboration behavior*.

The *business collaboration behavior* itself is an abstract concept. It is planned, that in future versions of UMM there might exist different approaches to describe the choreography of a *business collaboration*. Nevertheless in this version the only valid specialization is a *business collaboration protocol*. Therefore a *business choreography* is currently always defined by a *business collaboration protocol*. A *business collaboration protocol* itself is composed of zero to many *business collaboration activities* and of zero to many *business transaction activities*. At least one *business collaboration activity* or one *business transaction activity* must be present in a *business collaboration protocol*. The transitions which define the flow among the *business collaboration activities* and the *business transaction activities* can be guarded by the states of *business entities*.

A *business collaboration activity* is refined by another *business collaboration protocol*. Only the nested *business collaboration protocols* are refined *business collaboration activities*. Furthermore it is possible, that a *business collaboration protocol* is nested in different *business collaboration activities*. In the figure above this is represented by the (0..n) to 1 aggregation relationship between *business collaboration activity* and *business collaboration protocol*.

Finally a *business transaction activity* is refined by a *business transaction*. Because the *business transaction* is a concept which belongs to the *business interaction view*, it will be described in more detail in the next chapter. A *business transaction* must at least once be used to refine a *business transaction activity*. A *business transaction* can be nested in different *business transaction activities*. Therefore the aggregation relationship between *business transaction activity* and *business transaction* is (1..n) to 1.

The next view defines a choreography which leads to synchronized states of *business entities* on both sides of the interaction. It is the so called *business interaction view*.

Business Interaction View

This view contains exactly one *business interaction*, which leads to a synchronized business state between the two *authorized roles*, executing the

Reaching a synchronized business state

business interaction. Figure 8–12 shows the *business interaction view* at a glance.

Fig. 8–12 Overview of the business interaction view

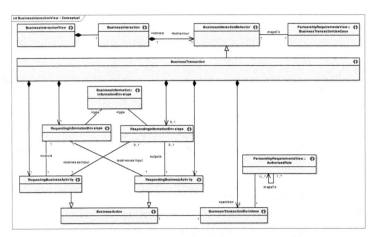

Furthermore a *business interaction* is a persistent representation of a synchronization of *business states* between *authorized roles*. The *business interaction behavior* describes the choreography of the synchronization and the required information exchanges. A *business interaction* is composed of exactly one *business interaction behavior*. The *business interaction behavior* is based on the requirements which have been defined in the corresponding *business transaction use case* of the *business requirements view*. Each *business transaction use case* is mapped to one *business interaction behavior* and each *business interaction behavior* is mapped to exactly one *business transaction use case*.

The stereotype *business interaction behavior* is an abstract concept. In a future version of UMM there may exist different approaches to describe the choreography and information exchanges in a *business interaction*. Nevertheless in this version the only specialization which is valid is the *business transaction*. The *business transaction* is an atomic *business process* which takes place between two roles. It involves sending business information from one role to the other. A reply is optional.

A *business transaction* is split up into partitions - each role has a partition. Hence a *business transaction* is composed of exactly two *business transaction swimlanes*, where each swimlane relates to one *authorized role*. Nevertheless an *authorized role* may be assigned to multiple *business transaction swimlanes* in different *business transactions* but to only one *business transaction swimlane* within a specific *business transaction*. This means, that the two swimlanes of a *business transaction* must be assigned to different roles.

Business transactions and the use of swimlanes

Within a *business transaction* each role performs one *business action*. The requesting role performs the *requesting business activity*, the responding role performs the *responding business activity*. Each *business action* is assigned to a swimlane and each swimlane contains exactly one *business action*. Each role can perform multiple *business actions* but only in different *business transactions*. A *business transaction* is composed of exactly one *requesting business activity* and exactly one *responding business activity*. Both business activities are specializations of a *business action*.

The output of a *requesting business activity* is a *requesting information envelope*. The *requesting information envelope* serves as an input for the *responding information activity*. The *responding information envelope*, which is the output of a *responding business activity* and the input of a *requesting business activity* is optional. Therefore a *business transaction* is composed of exactly one *requesting information envelope* and zero or one *responding information envelopes*. The *requesting information envelope* as well as the *responding information envelope* are instances of the type *information envelope*. A *requesting business activity* outputs exactly one *requesting information envelope* and a *requesting information envelope* is created by exactly one *requesting business activity*. A *requesting business activity* can receive zero or one *responding information envelopes* as input and a *responding information envelope* is input to exactly one *requesting business activity*. *Requestor*

The output of a *responding business activity* are zero or one *responding information envelopes* and a *responding information envelope* is created by exactly one *responding business activity*. The input of a *responding business activity* is exactly one *requesting information envelope* and a *requesting information envelope* is input to exactly one *responding business activity*. *Responder*

Both the *requesting information envelope* and the *responding information envelope* are stereotypes of the base class *object flow state*. The type of the object flow state is defined by the *information envelope* that is a stereotype of base class.. Multiple *object flow states* can be instances of the same *class*. For the *business transaction view* this means, that different *requesting information envelopes* or *responding information envelopes* might be instances of the same *information envelope*. An *information envelope* can therefore be reused in different *business transactions*.

The last subview within the *business transaction view* to be analyzed is the *business information view*.

Business Information View

The artifacts within the *business information view* describe the information which is exchanged in a *business interaction*. As already explained in the last paragraph, the *requesting information envelope* as well as the *respond-* *Describing the exchanged information ing information envelope* are of type *information envelope*. The information envelope serves as a cover for all the information exchanged between *requesting business activities* and *responding business activities* and vice versa. As Figure 8–13 shows, the information included in the envelope is structured by classes which are stereotyped as *information entity*.

Fig. 8–13 Overview of the business information view

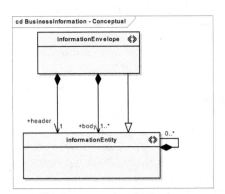

Information entities can be nested as the unary composition in the figure above shows. An *information envelope* contains a header and one ore more bodies. The header as well as the body are *information entities* themselves. Therefore an *information envelope* is composed of exactly one *information entity* with the role name *header* and one or more *information entities* with the role name *body*. The third generalization connector in the figure above shows, that an *information envelope* is a specialization of an *information entity*, which fulfills all the rules for the *information envelope* as well.

Not included in the current UMM foundation module are rules, which define how to build *information entities*. However current effort focuses on the modeling of *information entities* using the Core Component Technical Specification (CCTS) approach, which is discussed in more detail in chapter 10.

Following the description of the conceptual UMM meta model the next chapter will focus on the OCL constraints, which restrict the modeling in order to ensure a valid UMM model.

8.4   OCL constraints as the base for validation

As already mentioned in chapter 8.2.3, the base for the UMM validation are OCL constraints. The OCL constraints have been derived from the UMM meta model and are split up into four significant parts with additional sub parts.

Business collaboration model
Business domain view
Business requirements view
    Business process view
    Business entity view
    Partnership requirements view
Business transaction view
    Business choreography view
    Business interaction view
    Business information view

The structure of the OCL constraints

As one can see, the structure of the OCL constraints follows the UMM meta model. This is not arbitrary but organized by UN/CEFACT in order to facilitate maintenance and usage of the constraints. The OCL constraints within the *business collaboration model* part are applicable for the whole model. They ensure, that the model itself has a valid UMM structure. The following three main bullets indicate the constraints according to the three different main packages within the UMM. Furthermore the constraints within the *business requirements view* and the *business transaction view* are finer grained according to the subviews of the packages.

By using OCL constraints as the base for a validation engine errors can arise. The next chapter will focus on the process of transforming the OCL constraints into a validation engine and the problems, which might occur.

8.4.1   Validation techniques

After having specified the constraints on UMM in natural language in the specification they are transformed into OCL. OCL provides a common understandable base which guarantees, that no ambiguities occur. These OCL constraints must now be transformed into a validation engine which consists of three significant parts:

    a model to be validated
    a set of OCL constraints, which apply to the model
    a validation engine, which can validate the model against the OCL constraints

From the software engineers perspective three design approaches are possible. The next three paragraphs will give an overview of the three approaches and evaluate the pros and cons of each approach. Finally our method of choice will be presented.

Generic OCL validator

This approach can be considered as the most flexible one. The validator is implemented as stand-alone application or as a service. Figure 8–14 shows the architecture of a generic OCL validator. The input model which should be validated must be an XML Metadata Interchange (XMI) representation. Users can submit their models in the form of an XMI file to a stand-alone application or the XMI representation e.g. via Simple Object Access Protocol (SOAP) to a web service, which implements the validator service. Furthermore the validator requires the OCL constraints and the meta model as an input. The validator then checks the XMI document against the meta model and the OCL constraints.

Fig. 8–14 Generic OCL validator

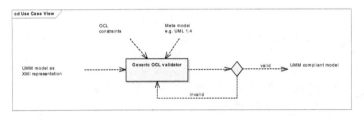

Although this approach is the most flexible one, in practice it has some shortcomings. It is difficult to represent a complicated UMM model in an unambiguous way in XMI. Especially the different modeling tools like Enterprise Architect, Poseidon or Rational Rose, just to name a few, do not export XMI in a congruent way - they use different XMI flavors.

The next shortcoming is the lack of usability for the modeler. When a UMM model is built from scratch, the modeler starts with the *business domain view*. After having finished this view, the user continues with designing the *business requirements view*. From the point of usability, it would be a great enhancement for the user, if he could validate the *business domain view* alone. When the *business domain view* is valid, the user then continues designing the *business requirements view* and so on. This bottom up validation approach requires a lot of validator invocations. Considering the fact, that for every modeling invocation the model has to be exported to XMI first and then transferred to a standalone application or a service, we argue, that the lack of efficiency makes this validation approach ineligible for UMM validation.

Another shortcoming of this approach is the presentation of error messages to the user. Clear and meaningful error messages help the modeler to correct the mistakes found by the validator. By using the generic approach, detailed error messages are not possible, which is a significant decrease in usability. Currently no mechanism in OCL is known, which could combine the constraints and personalized error messages.

However the chance to submit OCL constraints, a UML meta model and an arbitrary UMM model to a validation engine sounds tempting with regard to a generic, reusable validation engine. Nevertheless from the software engineer's perspective this approach is difficult in terms of feasibility. The difficulty in implementation and the enormous effort were the reasons, why this approach was not chosen.

OCL validator as an Add-In for a modeling tool

The OCL validator Add-In is implemented similarly to the generic OCL validator. While the generic validator is implemented as a stand-alone application or service this validation approach is realized as an Add-In for a modeling tool. In this example Enterprise Architect (EA) is chosen. As an input, the OCL interpreter requires the OCL constraints and a meta model. The user creates a UMM model with the modeling tool. Via a button inside the tool, the user starts a validation. Via the OCL interpreter the validation Add-In uses the OCL constraints and the meta model to validate the created model. Figure 8–15 shows the OCL interpreter Add-In at a glance.

Fig. 8–15 OCL validator Add-In

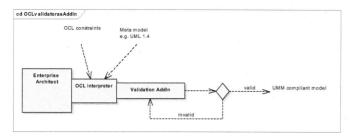

The advantage the OCL validator Add-In has over the generic OCL validator is the missing XMI representation. Because the XMI is missing, the ambiguities can be avoided and the performance yield allows bottom up validation runs. Enterprise Architect offers a well designed interface which can be used to access the elements of a model. The model representation which is offered by the interface is unambiguous.

Add-Ins for the Enterprise Architect are written in Delphi, C# or any other supported programming language. This means, that the software architect has to implement the OCL interpreter in one of the programming languages mentioned above. The disadvantage over the generic approach is apparent. A model which has been created e.g. with Rational Rose, can not be validated with this validation approach, as it is exclusively for Enterprise

Architect. Furthermore the implementation of an OCL interpreter as an Enterprise Architect Add-In is linked up with high development costs similar to the generic validator approach.

Validation Add-In for a modeling tool

The third approach for a validation engine is an Add-In for a modeling tool, where the OCL constraints are hard coded within the program code. Figure 8–16 gives an overview of this approach, which was the method of choice for this thesis. The validation Add-In represents the logic of the OCL constraints. Every constraint is taken and hard coded in C# code. The user starts the validation by pressing a button inside the EA. A major advantage over the first approach is the fact, that the user can validate sub models (e.g. only the business transaction view) as well. By providing this bottom up approach the user can step by step be guided towards a valid model.

Fig. 8–16 Validation Add In

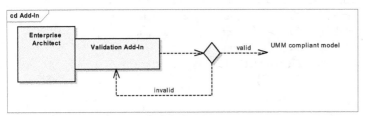

The first two methods were using an OCL interpreter, which parses the OCL constraints and validates a given model against them. The effort to write an OCL interpreter is quite high and the implementation e.g. via a web service often complicated or inefficient. Furthermore the performance of an OCL interpreter is slow and the error messages concerning the validated model are not fine grained.

Although the implementation cost for a validation Add-In must not be underestimated, it has one main advantage over the first two approaches. After implemented once, the validation Add-In has a very fast performance and provides the user with helpful error messages concerning the created model. Therefore this approach was our method of choice when we implemented the UMM validator. The design of our Add-In allows an easy extension and maintenance of the validation procedure, which helps adapting the validator to changes in the UMM standard.

Chapter 8.5 will focus on the Add-In in detail and analyze its advantages and disadvantages. Before we continue with the validator in detail we will analyze the transformation of an OCL constraint into the validator.

8.4.2   Transforming OCL constraints into a validation engine

In order to implement the correct logic in the validator, the OCL constraints have to be transformed into C# code. As an example for a transformation we will use a constraint which applies to the business collaboration model. Expressed in natural language the constraint states the following:

A *BusinessCollaborationModel* MUST NOT contain more than one *BusinessDomainView* package (but it MAY contain no *BusinessDomainView* package at all)

Natural language representation

Accordingly the OCL constraint is:

```
[250] package Model_Management
[251] context Model
[252]     inv zeroToOneBusinessDomainView:
[253]         self.isBusinessCollaborationModel() implies
[254]         self.ownedElement->select(isBusinessDomainView())->size()<=1
```

Listing 8–2 OCL constraint according to the natural language representation

Our goal is now transforming the OCL constraint into a C# code, which is then part of the validator. Every constraint is validated by using a method, which has the same name as the invariant of the OCL constraint. The name of the invariant in our example is zeroToOneBusinessDomainView.

The code representation of the OCL constraint zeroToOneBusinessDomainView is:

```
[255]     private bool checkOCL_zeroToOneBusinessDomainView() {
[256]         bool rv = false;
[257]         int c = 0;
[258]         EA.Package package = (EA.Package)repository.Models.GetAt(0);
[259]         foreach (EA.Package p in package.Packages) {
[260]             if(p.Element.Stereotype==
[261]                 UMM_Stereotype.BusinessDomainView.ToString())
[262]                 c++;
[263]         }
[264]         if ( !(c == 0 || c == 1)) {
[265]             this.validatorMessages.Error("Invalid number of
[266]             BusinessDomainView(s) detected.", new Constraint("A
[267]             BusinessCollaborationModel must not contain more than
[268]             one BusinessDomainView package (but it may contain
[269]             no BusinessDomainView package at all).",this,
[270]             new StackFrame()));
[271]             rv = true;
[272]         }
[273]         return rv;
[274]     }
```

Listing 8–3 C# code representation of the OCL constraint

As one can see, the name of the C# method equals the name of the OCL invariant with an additional checkOCL_ prefix. By applying this convention the maintenance of the constraints is facilitated, as the according method can easily be found within the code.

Every validation method for an OCL constraint returns a boolean value. In case one constraint is violated, the processing of the validation stops, and the user is presented the error message. The reason for doing this is because a lot of OCL constraints interrelate. For instance constraint B checks the actors within an use case diagram and constraint A checks if the use case diagram exists. If constraint A fails (because there is no appropriate use case diagram), it makes no sense to check constraint B. If no use case diagram is present, it makes no sense any more to check if actors are present.

Line 260-263 checks the occurrence of *business domain views* within the model. If the number is neither 0 nor 1, an error is raised. In line 265 an error is added to a collection of error messages. As we will see later, the collection of error messages is presented to the user after a validation run. We now continue with the validation engine.

8.5   The UMM validation Add-In

In this chapter we would like to explain the validation engine we implemented. As mentioned before, the validator among other features like the worksheet editor or a BPEL transformer is implemented as an Add-In within the Enterprise Architect.

The first issue to be discussed is the architecture of the validator. In the last chapter we already saw, how OCL constraints are transformed into C# methods. We now would like to examine how the validator is organized.

8.5.1   Architecture

The validator itself is split up into four major parts. Every major part of the UMM model namely *business domain view, business requirements view* and *business transaction view* as well as the *business collaboration model* (which is the model per se) has its own validator.

Figure 8–17 shows the UMM validator's class structure. For the sake of clarity within the class diagram less important attributes and operations have been left out.

Fig. 8–17 Class diagram
of the validator

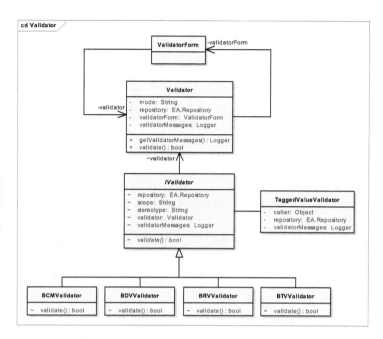

The class *ValidatorForm* represents the graphical user interface (GUI). Validation results are presented within the GUI and validation specific information is presented to the user. A validation specific information for instance is the scope, which the validator currently operates on or a progress bar, which shows the user the overall progress of a validation run. When talking about a scope within the validator, a specific subview within the model is meant, on which the validator operates. Scope can for instance be the *business domain view* or the *business information view* or the entire model just to name a few.

The class *Validator* holds the messages, which are generated by the different sub validation classes. Its method *getValidatorMessages()* is used by the class *ValidatorForm* to access the messages, which a validation run has generated. Furthermore it invokes the right sub validation routine depending on the current scope of the validator. The attribute *repository* is a reference to the current model which is opened in the Enterprise Architect. Via this reference every model element within the current model can be accessed.

The class *IValidator* is defined abstract with exactly one abstract method named *validate()*. Every subclass inherits the attributes of *IValidator* and must implement the abstract method validate(). The attribute *repository* as already mentioned holds a reference to the current model. The attri-

bute *validatorMessages* contains the messages, generated by the sub
validation routines. *Scope* as already mentioned defines the current scope of
the validator. In *scope* the package identification of the clicked package is
stored as an integer. The attribute *stereotype* holds the stereotype of the
scope. The attribute *validator* holds a reference to the class *Validator*. This
is necessary, because via that variable changes in the user interface can be
invoked. The most important change in the user interface during a validation run is the change of the progress bar and the change of the text in the
status bar.

The subclasses of the abstract class *IValidator* are responsible for validating the corresponding view in the UMM model whereas the first three
letters of the class name refer to the abbreviation of the according view. The
class *TaggedValueValidator* validates the tagged values within the current
scope of the validator.

The second issue to be discussed is the granularity of a validation. The
validator can operate in two different modes. The so called *scope validation*
(bottom up validation) validates a subview of the model e.g. the business
transaction view, whereas the *overall* (top down) *validation* validates the
whole model.

*Special tagged value vali-
dator*

8.5.2   Scope vs. overall validation

While designing a UMM model the user finishes one view after the other.
Normally one starts with the *business domain view* and its subviews, continues with the *business requirements view* and then finishes the *business
transaction view* and the corresponding subviews. After having for instance
finished the *business domain view*, the user wants to check, whether the created view is valid or not, before continuing with the *business requirements
view*. At this point the scope validation can be used.

The user is able to right click on any package in the tree view of the
Enterprise Architect. The tree view in the EA is the package tree on the right
hand side, which gives an overview about all created packages and their
hierarchical order. After having clicked on a package in the tree view, the
validator determines the stereotype of the clicked package and opens the
validator form. The scope of the validator is automatically set to the scope
of the package. Hence the user can start a validation run, which exclusively
validates the selected package. This leads to a great improvement in validator performance, as the user does not need to validate the whole model. Furthermore it enhances the usability, as only error messages, specific to the
selected package are presented.

A bottom up approach

In contrast to the scope specific validation, the overall validation validates the UMM model as a whole. This feature is used, after the user has fin-

A top down approach

ished an UMM model and wants to check its overall validity. Nevertheless the user can always initiate an overall validation during the design process of a model. However seen from the point of usability and performance an overall validation does only make sense, when the entire UMM model with its according subview is finished. Furthermore the overall validation functionality is used by the BPSS and BPEL transformer as well. Before a transformation is invoked, the transformer starts an overall validation to ensure a valid model. If the validator does not return any error messages, the transformer starts the transformation. Hence the transformer can anticipate a valid model.

We will now examine the validation of a specific package in detail.

8.5.3   Validation of a package - a deeper insight

As already mentioned at the beginning of chapter 8.4 the OCL constraints in the UMM specification are split up into four significant parts with additional sub parts. When leading over the OCL constraints into a validation engine, a programmer is facing certain problems, which we will try to analyse in this chapter.

At first, the structure, in which the OCL constraints are organized within the UMM specification is not suitable for direct transformation. As we saw in the last chapter, the chance to validate a specific subpackage must also be given in order to allow a scope specific validation. In scope specific validation, the user must have the chance to validate the following packages:

Business collaboration model
Business domain view
  Business area
  Process area
  Business category
Business requirements view
  Business entity view
  Business process view
  Collaboration realization view
  Collaboration requirements view
  Transaction requirements view
Business transaction view
  Business choreography view
  Business interaction view
  Business information view

Common packages within a UMM model

From the structure above two interesting facts can be derived. As first element the *business collaboration model* is mentioned. The *business collaboration model* represents the whole model. If the user initiates scope specific validation by clicking on the very top package in the tree view (which is the whole model), an overall validation is started. That means, that the scope of the validator is the whole model - therefore the validator operates as if it would be in overall mode.

Furthermore we can see, that the structure for scope validation is finer grained then the structure in which the OCL constraints are available from the documentation. E.g. in the UMM specification the OCL constraints of the *collaboration realization view, collaboration requirements view* and *transaction requirements view* are combined into the *partnership requirements view*.

Difference between OCL and UMM package structure granularity

The first exercise is to split up the OCL constraints and assign them to the correct subview as mentioned above. We then create a class for every view (as we saw in Figure 8–17) and furthermore create a method for every subview and a method for every OCL constraint. Figure 8–18 shows a class diagram for the *business transaction view* validator. In the upper right corner we can see the abstract class *IValidator* from which the *BTVValidator* inherits.

Fig. 8–18 The business transaction view validato

```
cd Validator

                                                                    IValidator
                              BTVValidator

  ~  BTVValidator(Logger, EA.Repository, String, String, Validator)
  -  check_BusinessChoreographyView() : bool
  -  check_BusinessInformationView() : bool
  -  check_BusinessInteractionView() : bool
  -  check_BusinessTransactionView() : bool
  -  checkGeneralConstraintsOnBusinessInteractionView() : bool
  -  checkOCL_AllowedElementsInBusinessInformationView() : bool
  -  checkOCL_AllowedModelElementsInBCP() : bool
  -  checkOCL_BCArefinedByExactlyOneBCP() : bool
  -  checkOCL_BCdescribedByOneBusinessChoreographyBehaviour() : bool
  -  checkOCL_BCPmapsToBCUseCase() : bool
  -  checkOCL_BCVcontainsExcactlyOneBC() : bool
  -  checkOCL_BehaviorOfBIdescribedByExactlyOneBusinessInteractionBehavior() : bool
  -  checkOCL_BIBmapsToExactlyOneBusinessTransactionUseCase() : bool
  -  checkOCL_BIVcontainsExactlyOneBI() : bool
  -  checkOCL_BTArefinedByExcactlyOneBT() : bool
  -  checkOCL_BusinessTransactionHasExactlyTwoBTSwimlanes() : bool
  -  checkOCL_BusinessTransactionSwimlaneClassifier() : bool
  -  checkOCL_contentsOfInformationEntity() : bool
  -  checkOCL_ContentsOfRequestingPartition() : bool
  -  checkOCL_ContentsOfResponderPartition() : bool
  -  checkOCL_InformationEnvelopeHasBodies() : bool
  -  checkOCL_InformationEnvelopeHasHeader() : bool
  -  checkOCL_ObjectFlowStateHasClassifier() : bool
  -  checkOCL_packagesAllowedInBTV(String) : bool
  -  checkOCL_TrInitialState2RequestingBusinessActivity() : bool
  -  checkOCL_TrPossibleRespondingInformationEnvelope2RequestingBusinessActivity() : bool
  -  checkOCL_TrRequestingBusinessActivity2FinalState() : bool
  -  checkOCL_TrRequestingBusinessActivity2ReqInfEnvelope() : bool
  -  checkOCL_TrRequestingInformationEnvelope2RespondingBusinessActivity() : bool
  -  checkOCL_TrRespondingBusinessActivity2RespondingInformationEnvelope() : bool
  -  checkOCL_TrRespondingInformationEnvelope2RequestingBusinessActivity() : bool
  ~  validate() : bool
```

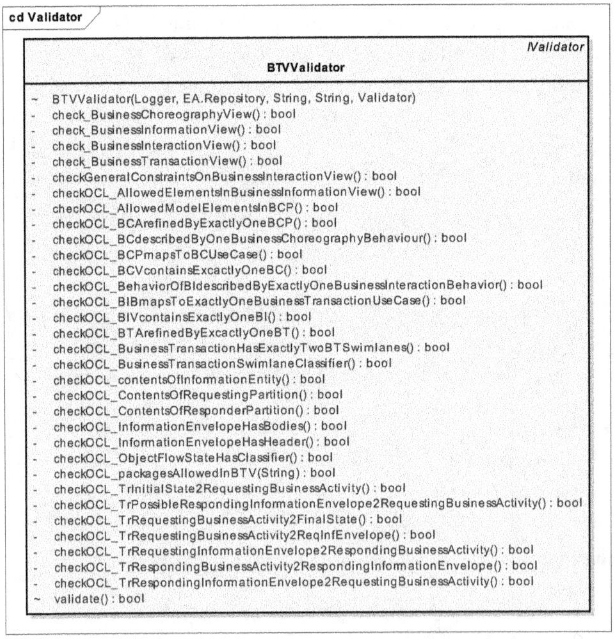

The first method of the class is the constructor which receives five parameters which initialize the *BTVValidator*. The last method *validate()* is inherited from the superclass *Validator*. By using the scope information, the method *validate()* invokes the correct method for the required subview. Four methods are responsible for validating the correct subview:

check_BusinessTransactionView()
check_BusinessChoreographyView()
check_BusinessInteractionView()
check_BusinessInformationView()

One sub-routine for every subview

The method *check_BusinessTransactionView()* validates the whole *business transaction view*. This is done by one after the other invoking the three other methods *check_BusinessChoreographyView()*, *check_BusinessInteractionView()* and *check_BusinessInformationView()*.

As an example we now take the *check_BusinessInformationView()* and examine how the correct OCL constraints are validated within this method. The method is implemented as following:

```
[275] private bool check_BusinessInformationView() {
[276]
[277]     validator.incrementValidatorProgress();
[278]
[279]     bool error = checkOCL_AllowedElementsInBusinessInformationView();
[280]     if (!error)
[281]         checkOCL_InformationEnvelopeHasHeader();
[282]     if (!error)
[283]         error = checkOCL_InformationEnvelopeHasBodies();
[284]     if (!error)
[285]         error = checkOCL_contentsOfInformationEntity();
[286]
[287]     validator.incrementValidatorProgress();
[288]
[289]     //Validate the TaggedValues
[290]     if (!error)
[291]         new TaggedValueValidator(this.validatorMessages,
[292]         this.repository,this).validatePackageAndContentTV(this.scope);
[293]     return error;
[294] }
```

Listing 8–4 Invocation of a business information view validation

For the *business information view* four OCL constraints are relevant, which are validated in the lines 279, 281, 283 and 285. The validation of a specific OCL constraint is done by invoking the correct sub method e.g. *checkOCL_AllowedElementsInBusinessInformationView()*. Every method which validates an OCL constraints returns true in case an error occurred or false instead. In case one method returns an error, the validation stops and the user is presented the error message. This is due to the fact, that the OCL constraints and therefore the methods validating the according constraints interrelate. If for instance the method in line 283 returns true because an error occurred while checking if the *information envelope* has message bodies, it does not make sense to invoke the next method in line 285. If the *information envelope* is missing the message bodies it is useless to check the contents of the message bodies.

Line 291 invokes the validator, which validates the tagged values of the *business information view*. Line 277 and 287 increment the progress bar in the user interface.

In the example above we saw how a specific subview validation for the *business information view* is implemented. We now want to look at the way, the *BTVValidator* invokes the methods for the validation for the corresponding subviews. The following example shows the method *check_BusinessTransactionView()*, which invokes the methods, which validates the whole *business transaction view*. For the sake of clearness some parts have been shortened. Three dots indicate, that parts have been omitted.

```
[295] private bool check_BusinessTransactionView() {
[296]        EA.Package p = this.repository.GetPackageByID(Int32.Parse(scope));
[297]        bool error = false;
[298]        ArrayList packages = new CustomPackage(...).getPackages();
[299]        foreach (CustomPackage cp in packages) {
[300]                EA.Package pa = cp.Package;
[301]                this.scope = pa.PackageID.ToString();
[302]                String stereotype = pa.Element.Stereotype;
[303]                if (stereotype == ...BusinessChoreographyView.ToString())
[304]                        error = check_BusinessChoreographyView();
[305]                else if (stereotype == ...BusinessInteractionView.ToString())
[306]                        error = check_BusinessInteractionView();
[307]                else if (stereotype == ...BusinessInformationView.ToString())
[308]                        error = check_BusinessInformationView();
[309]                else {
[310]                        error = true;
[311]                        this.validatorMessages.Error(...);
[312]                }
[313]                if (error)
[314]                        break;
[315]        }
[316]        if (!error)
[317]                new TaggedValueValidator(...);
[318]        return error;
[319] }
```

Listing 8–5 Invocation of a business transaction view validation

In line 296 the package is retrieved using the current scope. In this case the package is the *business transaction view*. Line 298 retrieves the packages within the *business transaction view* and puts them in the correct order.

Why a sort is necessary is explained in the next paragraph. In line 299 to 315 the method iterates over the found packages and invokes depending on the stereotype of the package the correct subvalidation routine. In case one of the subpackages is not valid the *error* flag is set to true and the validation stops. If no error occurred, the tagged value validator is invoked in line 317.

The method *check_BusinessTransactionView()* itself returns a boolean value as well. The purpose is, that if the whole model has to be validated one after the other the *business domain view*, the *business requirements view* and the *business transaction view* validators are invoked. If an error occurs within the *business domain view*, the *BDVValidator* returns true and the validators for the *business requirements view* and the *business transaction view* are not invoked.

The other sub validators, *BCMValidator*, *BDVValidator* and *BRVValidator* are implemented analogously. By using the approach above both, the scope validation and the overall validation can be executed by the same validator. In case of an overall validation the *BCMValidator* is invoked, which then invokes *BDVValidator*, *BRVValidator* and the *BTVValidator*. Hence, the whole model is validated.

We will now have a deeper insight into the mechanism which is responsible for the validation of the tagged values.

8.5.4   Tagged value validator

As already mentioned at the beginning of this chapter, tagged values play an important role within the UMM standard. The presence of tagged values and their correct values are crucial for applications using the UMM model. The tagged value validator is invoked once for every subview of the UMM model. In case we would for instance validate the *business transaction view*, the validator would be called for every subview namely:

Business transaction view
  Business choreography view
  Business interaction view
  Business information view

Listing 8–5 shows the C# code for the validation of a *business transaction view*. In line 317 the tagged value validator for the tagged values of the *business transaction view* is invoked. Please note, that the invocation in line 317 only validates the tagged values of the *business transaction view* itself. The tagged values of the subpackages are validated in the according subroutines which are invoked in the lines 304, 306 and 308. This mechanism is necessary in order to allow a package specific validation as needed for the bottom up validation approach.

Figure 8–19 shows an overview of the tagged value validator. Every sub routine of the validator invokes the tagged value validator by calling the *validatePackageAndContentTV()* method and passing the scope variable to it. Within the scope variable the information is stored, on which package the tagged value validator is supposed to operate. The method *validatePackageAndContentTV()* then in turn determines the correct package from EA's package collection and validates its elements.

Fig. 8–19 *Class diagram of the tagged value validator*

For every specific subpackage a method is provided, which checks the occurrence and validity of the corresponding tagged values. In general we distinguish between four different data types which a tagged value can have, namely *string, boolean, duration* and *integer.* The tagged value validator not only checks if a specific tagged value is present but also if the tagged value has an appropriate data type. In case either the tagged value is missing or its data type is wrong, an error is raised and added to the collection of error messages.

In contrast to the regular validator, which validates the UMM model, the tagged value validator does not stop the validation if an invalid or missing tagged value is detected. This is only possible because we do not have an interrelation between the tagged values as we have between the constraints which apply to the UMM model.

8.5.5   Difficulties which accompany the validation

One major problem which a programmer is facing is due to the implementation of the interface which is provided by Enterprise Architect. Internally in Enterprise Architect packages are identified using a unique ID. As we have seen in the chapter about scope vs. overall validation, packages are essential to the validator. Therefore also internally in the validation Add-In packages are identified using the Enterprise Architect internal ID. When validating for instance the *business transaction view* (as shown above), the order in which the subpackages are validated is important.

First the *business choreography views* must be validated followed by the *business interaction views* and the *business information views*. For the validation per se the order is not that crucial as for the usability. Errors occurring in the *business choreography view* should be presented first, then the errors of the *business interaction view* and so on. Unfortunately the Enterprise Architect internal IDs are not necessarily ordered as the underlying data structure is a relational database. When validating the *business transaction view*, the programmer has to first extract all subpackages of the view and sort them - first *business choreography views*, then *business interaction views* and then *business information views*. For a beginner to the Add-In programming with Enterprise Architect this is a major stumbling block, because the implementation anomaly in the Enterprise Architect is not immediately apparent.

EA does not store packages in correct order

The second issue is more an exception than a difficulty. The *business domain view* has a special behavior as its elements can be nested recursively. Therefore the programmer first has to extract all relevant views recursively, check their consistency and then invoke the validation.

Another issue which a programmer is facing is a performance problem which can occur especially when a UMM model gets very large. Figure 8–20 shows the UMM Add-In and its context within Enterprise Architect. Because Enterprise Architect is build on a relational database, namely Microsoft Access every access to a data collection of the EA interface requires a select query on the database.

Fig. 8–20 The EA architecture at a glance

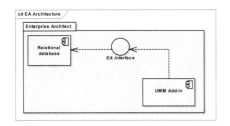

If the validator is for instance performing a validation of the whole model, every package and its elements have to be retrieved from the database. This can take a while and reduces the overall performance of the validator. One could argue, that a one-time transformation of the model into a C# specific data structure would ease the performance as a C# specific data structure would be very fast especially for searching and evaluating elements. However we must not forget, that the validation is performed several times while modeling and therefore we would have to transform the model into the C# specific data structure for every validation run. Hence the performance gain is lost. However for specific appliances like a BPSS transformation the conversion of the model into a C# specific data structure does make sense. A transformation is a one-time action and therefore a performance gain would be possible.

In the next chapter we will have a brief look on the way, validation results are presented to the user. We will see, that meaningful error messages are a key issue to a good usability of the validator.

8.5.6   Presentation of validation results

We will now focus on the presentation of the validation results to the user. A clear and easy conceivable user interface together with meaningful error messages help the user to model a valid UMM model. Whenever a validation is invoked, a new validator window will pop up and present the user the validator user interface. Figure 8–21 shows the interface before a validation run.

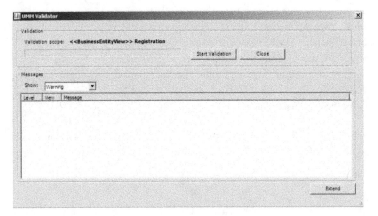

Fig. 8–21 The validator user interface

The field *Validation scope* indicates the current scope of the validator. In the example shown above the validator currently operates on a *business entity*

view named *registration*. The stereotype of the selected package is shown
between the guillemets. By clicking the *Start Validation* button the user
commences a validation run - in this case a validation on the package
denoted in the *validation scope* field. The messages which are generated by
the validator will be presented to the user in the messages pane. The
progress of the validation is shown by the progress bar under the validation
scope. In Figure 8–22 we can see the validator user interface after a validation run on the entire model.

*Fig. 8–22 The user
interface after a
validation run*

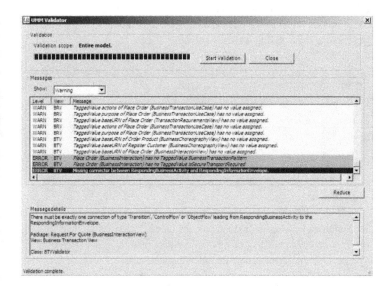

The status text in the lower left corner and the finished progress bar indicate, that the validation run is complete. The generated messages are shown
to the user in the messages pane. Within the validator we distinguish
between four different error levels. The error levels ascending corresponding to their severity are: *INFO, WARN, ERROR* and *FATAL*.

INFO messages are used to inform the user about incidents in the model
which are not wrong in the sense of the meta model and the OCL constraints. If for instance no *business entity view* is present within the *business
requirements view* an error message with the level *INFO* is generated.
Although not wrong according to the meta model, the user can just have
simply forgotten to create a *business entity view*. The *INFO* message
reminds him of the missing *business entity view*. Within the message pane
INFO messages have a white background.

WARN messages are used to inform the user about incidents in the
model which are not mandatory but strongly recommended. If for instance a
tagged value is present but its value is missing, an error message with the

Error message distinction

level *WARN* is generated. Within the message pane *WARN* messages have a yellow background.

ERROR messages are used to inform the user about incidents in the model which are wrong in regard to the OCL constraints and hence to the meta model. In Figure 8–22 three *ERROR* messages are shown with a red background. The color distinction helps the user to easily differentiate between the error levels.

FATAL messages only occur, when a crucial validator internal exception is thrown. Although coded with greatest carefulness a one hundred percent correctness can not be guaranteed. Hence if a *FATAL* message occurs the user is automatically presented a small dialog which is shown in Figure 8–23.

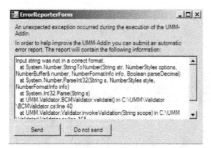

Fig. 8–23 The automatic error reporter

Similar to the system implemented by Microsoft Windows, the user is given the facility to send an error report to the UMM Add-In development team in case an unexpected error occurs. The error report includes the stack trace and the version numbers of the .NET Framework, the operating system, the UMM Add-In and the Enterprise Architects version number. Insofar the development team is able to quickly respond to reported errors and include a bug fix in the next release.

"E.T. phone home" functionality

By using the choice box shown in Figure 8–22, the user can select the level of the error messages to be shown in the message pane. The user is given the room to choose between three different levels namely *All*, *Warning* and *Error*. If *All* is choosen, the user is presented all error messages. In case *Warning* is choosen, the user is presented only messages with *WARN*, *ERROR* and *FATAL* level. If *Error* is choosen, the user is presented only *ERROR* and *FATAL* messages. Most modelers will presumably be interested in the *WARN*, *ERROR* and *FATAL* messages. Therefore these messages are presented by default.

If a user double clicks on a message within the message pane, the window extends and a detail section is shown. Within the detail section further information concerning the error message is given. If for instance a package like the *business domain view* contains invalid elements, the names of these elements are presented in the detail section. Furthermore the view and the specific package is shown as well. This allows the modeler to quickly navigate to the mistake within the model. As we can see in Figure 8–22, the user has clicked on an *ERROR* message which indicates a missing connector between a *responding business activity* and a *requesting information envelope*. The detail section gives the user a more specific description of the error and helps him to fix the mistake. Furthermore the package and view of the error occurrence is shown as well. In our case the missing connector can be found in the *business transaction view's* package *request for quote* which is of type *business interaction view*. For a future version a new feature within the validator is planned, which allows the user to quickly navigate to the error in the model just by clicking a button in the validators user interface.

Error messages which occur because of a missing or wrong tagged value are presented to the user in an italic font. This facilitates the distinction between an error message caused by for instance a misplaced modeling element and a missing tagged value.

In the next chapter we will look at the requirements of a BPSS validation which are a slightly tighter than the regular UMM requirements. Hence the validator routine has to be stricter as well.

8.6   The need for a special BPSS validation

One significant feature of the UMM Add-In is the BPSS/BPEL transformer, which allows the user to generate a Business Process Specification Scheme (BPSS) instance from the UMM model. Furthermore a Business Process Execution Language (BPEL) file can be generated as well.

    If one wants to generate a BPSS from a UMM model, two additional constraints must apply to the UMM model in order to guarantee a faultless functionality of the BPSS transformer. By passing a boolean variable to the constructor of the validator, a BPSS specific validation run can be invoked. We will now examine the additional requirements which apply to a BPSS validator. Figure 8–24 shows a sample *business collaboration protocol* which consists of an initial state, two final states and a *business transaction-activity*. The example is valid against the UMM meta model. However for BPSS additional restrictions apply. In order to guarantee a faultless BPSS transformation, one final state must be named *failure* and the other one *success*. The BPSS validator takes into account the additional constraints. If the *business collaboration protocol* in Figure 8–24 would be validated, two errors would occur. Namely the two falsely named final states.

Additional constraints apply for BPSS and BPEL transformations

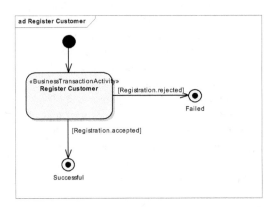

*Fig. 8–24 A sample
business collaboration
protocol*

Whereas the last constraint applied to the *business transaction view* the next one can be found in the *business requirements view*. Figure 8–25 shows a part of a *collaboration requirements view*. It depicts a *business collaboration use case* named *order product* with three parties participating in it. A collaboration with two participating *authorized roles* is called a binary collaboration. A collaboration with more than two participating *authorized roles* is called a multiparty collaboration - like the one we can see in Figure 8–25. In UMM a *business collaboration use case* must have at least two *authorized roles* which participate in the collaboration - nevertheless more participants are possible. However for the transformation into BPSS we have to consider additional constraints.

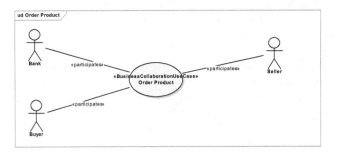

*Fig. 8–25 A part of a
collaboration
requirements view*

The BPSS transformer of the current UMM Add-In supports the BPSS 1.10 standard. This BPSS standard does only support binary collaborations, multiparty collaborations are considered as deprecated. Hence before a BPSS transformation the model is checked for multiparty collaborations. In case collaborations with more than two participants are found, an error is raised.

Future releases of the UMM Add-In will support the generation of BPSS 2.0 as well. The BPSS 2.0 schema has an additional complex type called *MultiPartyCollaborationType* which allows the representation of a multiparty collaboration. Further information on the topic of multiparty collaborations and BPSS can be found at [EBB05].

After having worked with Enterprise Architect Add-In development for nearly one year, some shortcomings have became apparent. Either these shortcomings were or are due to missing implementations in EA or due to an ineptly implemented interface. In the next chapter we will center on the shortcomings of Enterprise Architect and focus on their implications on the modeling and validation procedure.

8.7   Shortcomings of the Enterprise Architect

Although the Enterprise Architect has proven to be a powerful and efficient modeling tool for UMM we have encountered some shortcomings as well.

Within the UMM standard UML *partitions* are widely used. In the *business process view* for instance, *partitions* denote the different *business partners* which participate in a shared *business process*. As we already know from chapter 8.3.2 the *business process activity model* may also denote important states of *business entities* that are manipulated during the execution of a business process. Business entity states which are communicated to a *business partner* are so called *shared business entity states*. The modeler therefore places the *shared business entity state* between the two *partitions* in a way that the left and right side of the shared business entity element touches the two *partitions* between which the *shared business entity state* is exchanged. Meanwhile the modeler can see the *partitions* and its elements underneath in the treeview of the Enterprise Architect. If an element is placed between two *partitions* is must not be arranged underneath a specific *partition* element in the treeview but on the same level as the *partition* elements. However Enterprise Architect often places elements which are exactly between two *partitions* under one of these *partitions*. This is wrong however the flaw is not immediately apparent to the modeler. After running a validation, the validator reports the error. The user then has to correct the position of the misplaced element in the treeview of Enterprise Architect. This issue can be regarded as an unhandily implemented function of EA because the program does not react properly on the actions the user takes in the drawing canvas.

Minor failures occur when using partitions

Another shortcoming also involving *partitions* occurs when the modeler uses patterns. The UMM Add-In offers six different patterns which represent the most common *business interactions*. Namely these patterns are *commercial transaction, information distribution, notification,*

Pattern implementation is not flawless

query/response, request/confirm, request/response. The workflow for creating a *business interaction view* includes the creation of a package and a activity graph. The modeler then drag and drops the chosen UMM pattern onto the *activity graph,* where the *business transaction* is created. All UMM patterns for a *business transaction* use partitions in order to distinguish between the requesting and the responding business partner. The elements of the different *partitions* are arranged in the treeview of the Enterprise Architect. However after dragging and dropping a business transaction pattern the elements of the different partitions are not shown underneath the according *partition* but all on the same level. This is apparently wrong and a validation run returns the flaw accordingly. However after a manual reload of the model, the elements of the *partitions* appear under the correct *partition.* This shortcoming can be considered as a rather minor one, however it restrains the usability.

Another problem which comes with Enterprise Architect is its support for UML 2.0. The UMM specification is build on the UML 1.4 standard and therefore requires UML 1.4 compliant models. With the introduction of UML 2.0 some diagram types and elements have encountered major and minor revision. The *State machine* was extended by a *protocol state machine* and also the specification for the *state machine* itself was renewed. Significant changes were also made in the *activity diagram* (*activity graph*) specification. Nevertheless the main problem is not the introduction of a new UML 2.0 standard but the users ability to use new features within Enterprise Architect. Although correct in regard to the 2.0 specification, some elements are simply wrong or not known within the 1.4 standard. The validator still guides the user towards a valid UMM model, however on his way to a valid model the user has the chance to use UML 2.0 features and therefore he is seduced to make mistakes. For a detailed distinction between the UML 1.4 [UMa04] and UML 2.0 [UMb04] standard we would like to refer the interested reader to the respective standards. Future version of the UMM standard will presumably also support UML 2.0 features and the shortcoming mentioned above will vanish.

UMM and UML 2.0 together show their foibles

Even though the shortcomings mentioned in the last paragraphs have proven to be bothersome for the development and modeling process, Enterprise Architect as a whole can be regarded as sophisticated and efficient modeling tool which serves the purpose of a UMM compliant modeling tool very well.

8.8   Conclusion and outlook

The last chapters have shown, that the UMM validator is a great enhancement to the modeler as it facilitates the creation of UMM compliant models.

Furthermore it lowers the threshold for inexperienced modelers to start with UMM modeling. However improvements of the validator are still possible.

Future versions could allow OCL constraints to be directly submitted to the validator, not for every validation run but only once and then be valid until new OCL constraints are submitted. The same holds for a dynamic meta model. As already mentioned in the chapter about validator architecture these approaches would require a great effort in regard to human resources. However such a generic approach would make the adaptation of the validator to changes in the UMM standard more flexible.

Furthermore an AI based system could provide the user with intelligent correction recommendations in case an error is reported by the validator. One could even go further and state, that an AI based system could autonomously correct minor mistakes. In order to help the UMM standard become more common in the business process modeling world an easy accessible and efficient validation should be established. Future implementations of the validator could be implemented as a web service and be accessible free of charge for interested modelers. Models would then be submitted by using XMI representations. In order to make this approach feasible, a commonly used XMI standard would be necessary. It is now up to the vendors of UML modeling software to comply with the XMI standards as at the moment to many specific flavors of XMI exist. This makes a generic, web service based UMM validation impossible.

9   Generating Process Specifications from UMM Models

Modern software development approaches tend to apply service oriented architectures. In a service oriented environment functionality and business logic is distributed over a network. Unlike monolithic software systems service oriented architectures provide lightweight information systems by reusing already available functionality. In this respect functionality is exposed as a service in a networked environment, but not bound to a specific information system. Hence, state of the art software systems follow the thoughts of software reuse by composing new services on top of already existing ones. However, a composite service requires to choreograph component services in order to realize the desired functionality.

Similar to general modern software systems, B2B information systems expose interfaces to their business logic to the outside world. An interface of a B2B system corresponds usually to one or more services allowing to conduct public business process. The service oriented approach provides potential business partners a facile and apace way to conduct business with each other. However, real world business processes are complex. Inasmuch they typically consist of more than one (request/response based) interaction between two partners. Thus, in order to execute real-world business processes we need to choreograph interactions between such business services interfaces.

Electronic business requires choreographing B2B systems

We have already outlined in this paper, that UMM is a well-accepted method to define B2B choreographies. UMM specifies business choreographies on a conceptual, model-based level, but not in a system-interpretable manner. Recently, several XML-based choreography languages have emerged in the field of service oriented architectures. They provide means for describing a process in order to configure compatible information systems. The specified processes are in turn executed and monitored by the information systems. Hence, there is a need to derive such machine-executable process specifications from graphical UMM choreographies. This closes the gap between business process models and their machine-execut-

Mapping UMM to machine-executable process specifications is desired

able process descriptions. Inasmuch this potentiates UMM to be the cornerstone of a model driven approach for B2B.

In the remainder of this section we illustrate the derivation of process specifications from UMM to the Business Process Execution Language (BPEL) [BEA03]. Furthermore we discuss the implementation of the derivation process as part of the UMM Add-In.

The Business Process Specification Schema (BPSS) [BPS03] has been developed as system-executable subset of the UMM. Hence, in our UMM Add-In we have additionally implemented a transformation engine that generates BPSS artifacts. Our implementation follows the approach described in [Mic06]. However, in order to avoid describing the same approach twice, we will only detail the mapping from UMM models to BPEL in this thesis.

9.1 Deriving BPEL processes from UMM choreographies

9.1.1 What is BPEL?

The Business Process Execution Language (BPEL) seems to be the most adopted amongst the several service choreography languages. It supersedes XLANG [XLA01] and the Web Service Flow Language (WSFL) [WSL01] by combining these two, formerly competitive standards. BPEL is considered as part of the web service stack. It choreographs web services defined by the means of the Web Service Description Languages (WSDL) [WSD01] in exchanging messages via the Simple Object Access Protocol (SOAP) [SOA03].

BPEL provides means to choreograph the behavior of a B2B business process. A business process is (most likely) a long running transaction spanning over several interactions between its participants. Business partners who participate in a business process might provide services to the process and consume other partner's services. BPEL specifies the execution order of such service calls, maps service calls to concrete web services and collates service providing and service consuming to participants according to the process semantic. Hence, every business partner knows the process interfaces of his partners as well as the flow of service calls. However, BPEL describes a business process from a particular partner's point of view. Let's consider a complex B2B scenario: if each participant specifies the same process just from his own point of view, the resulting process specifications describing the same process will most likely not match. Hence, we require an approach that specifies a process from a global view in order to derive matching process specifications for each participating business partner.

BPEL describes a process from a partner's view

UMM in conjunction with a set of transformation rules for BPEL (e.g. as proposed by [HH04]) conforms to such an approach.

BPEL describes two types of processes - *executable processes* and *abstract processes*. *Executable processes* are expected to be executed and monitored within a workflow system. When defining such process descriptions we must specify the structure of exchanged documents as well as branching conditions, which are evaluated at runtime. In addition, BPEL defines the execution flow of a business process using the concept of an *abstract process*. Such an *abstract process* - also called *business protocol* - is described by a subset of BPEL and might be used by collaborating partners to get a common, simple overview about the same process. *Business protocols* abstract details that are not necessary to comprehend the process workflow. Such details include the structure of exchanged messages, business rules determining branch selection or complex data manipulation steps. *Business protocols* are not expected to be executable and deterministic. However, *abstract processes* might be easily extended to gain executable process specifications.

abstract vs. executable BPEL processes

UMM defines no service binding layer in its current version. This means, that we currently have no chance to define concrete bindings to protocols and endpoints. Furthermore, describing business information in a reusable way by means of *Core Components* and generating document schemes thereof is currently under development. As a result of this development a UMM specialization module for *Core Components* will be released soon. Chapter 10 of this thesis deals with the mapping of business information to certain document formats. At the moment, the generation of BPEL descriptions based on UMM models is limited to *abstract processes*.

UMM currently supports the generation of abstract BPEL processes only

9.2 Implementing a BPEL transformation algorithm

A derivation of BPEL process descriptions from UMM compliant *business collaboration models* has already been defined on a conceptual level in [HH04]. The BPEL transformation engine implemented as part of the UMM Add-In is mostly based on mapping rules proposed in this paper.

9.2.1 Initiation of the transformation process

In order to initiate the generation we need to input at least one *business collaboration protocol* represented by an API object as outlined in chapter 4. One *business collaboration protocol* corresponds to exactly one BPEL process.

9.2.2   Identification of involved roles

In order to create partner-specific BPEL descriptions we foremost need to determine the participants of a business collaboration. In order to identify process roles, we retrieve the *business collaboration use case* that is associated with the corresponding *business collaboration protocol*. According to the UMM meta model, the *business collaboration use case* determines the roles participating in a collaboration.

In order to avoid operating on the generic *Enterprise Architect* API built for accessing arbitrary UML diagrams and model elements (as introduced in chapter 4), we first of all create a supportive data structure consisting of elements that are relevant to the transformation process. Figure 9–1 gives a conceptual overview of this data structure. By using the data structure - further called *collaboration role data structure* - we resolve complex UMM relationships once at the beginning of the transformation process. In other words, during the following transformation stages we have convenient access on the particular elements.

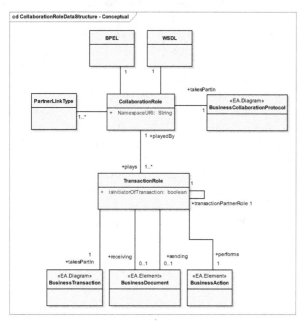

Fig. 9–1 Collaboration role data structure (conceptual)

The *collaboration role* class is the core element of the supportive *collaboration role data structure* (see Figure 9–1). A *collaboration role* represents an *authorized role* participating in a business collaboration. Since there is no need in the further transformation to access the *authorized role* API element directly, the data structure interface provides no reference to it.

However, the *authorized role* API element is used for the internal processing of roles.

Three types of artifacts - *BPEL processes*, *WSDL descriptions* and *partner link types* are created for each *collaboration role*. For each party a particular role is collaborating with, a *partner link type* is specified. Considering a multiparty collaboration, one *partner link type* is created for each
binary conversational relationship. If only two roles collaborate, each *collaboration role* has exactly one *partner link type* associated. Otherwise we
need more than one *partner link type*: one for each pair of business partners
interacting in the multiparty collaboration.

A partner's services are specified by the means of WSDL. In other
words, we use one WSDL instance to capture the *operations* a *collaboration
role* must provide as well as the *messages* that are exchanged as input or
output of these *operations*. Moreover, we use exactly one BPEL instance to
describe the flow of a partner's business process. BPEL choreographs activities within a business process by referencing to WSDL descriptions. In
UMM we specify the choreography of a collaborative business process by
means of a *business collaboration protocol*. It follows, that the choreography of a *business collaboration protocol* maps to a BPEL process. A *collaboration role* is associated with exactly one API element representing the
business collaboration protocol. The API element provides the required
information to construct the BPEL process.

In the UMM Add-In implementation we facilitated XML data binding
mechanisms in order to construct BPEL and WSDL artifacts. In the case of
BPEL we created a corresponding object structure using the *xsd tool* provided by the *.NET framework*. The *xsd tool* takes a W3C XML Schema
[XSD04] file as input and constructs an adequate object structure. Furthermore, it provides serialization and deserialization functionality along with
the object structure. Concerning WSDL we utilized the *service description*
class of the *.NET framework*. A *service description* instance corresponds to
a WSDL instance and is serializable as well. In addition the *service description* class provides means to extend a WSDL description as defined by the
specification. By means of the *extension* class we add *partner link types* to
WSDL descriptions as described by the BPEL specification.

In order to specify namespace declarations within BPEL and WSDL
artifacts each *collaboration role* has an associated *namespace URI*. This
URI should be defined by the user before the transformation is started.

*Artifacts for each collabo-
ration role*

*Relationship between BPEL
and WSDL*

*XML data binding supports
the constructing of BPEL
and WSDL artifacts*

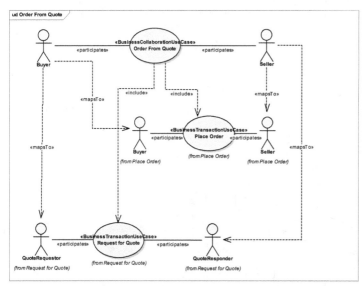

Fig. 9–2 Order from quote example: The collaboration role data structure resolves the mapsTo relationships defined in the collaboration requirements view

Another functionality of the *collaboration role data structure* is the handling of roles. At the beginning we once identify which collaboration role plays which role in a *business transaction*. As already outlined in chapter 7.2.3, the UMM *collaboration requirements view* is used to collate roles participating in a *business transaction* to roles of a *business collaboration protocol* (see Figure 9–2 for the *order from quote* example). We define such a relationship via a *mapsTo* dependency. With respect to the construction of the data structure, we resolve these relationships once at the beginning for every collaboration role. For every *mapsTo* leading from a particular collaboration role to a role participating in a transaction, we construct a corresponding *transaction role* object. A *transaction role* is double linked with its belonging *collaboration role*. In other words, with help of the data structure we can resolve the collaboration role that plays a certain transaction role as well as vice versa.

The data structure collates transaction roles to collaboration roles

A *transaction role* is further associated with the *business transaction* it takes part in. The *activity graph* representing the *business transaction* is again an API element of *Enterprise Architect's Automation Interface*. For convenience we determine if the particular role is the initiator of the *business transaction* or not. This is done by checking if a *requesting* or *responding business activity* is located in the *partition* that belongs to the role. Moreover, we retrieve the particular *requesting* or *responding business activity* as well as the business documents that are sent and received. This avoids resolving the same relationships multiple times again later on.

A transaction role takes part in a business transaction

Another useful feature provided by the data structure is the bi-directional association of roles, which interact in a transaction. By means of partner role interconnection in conjunction with the double link between a *transaction role* and its *collaboration role*, we determine which participants in a collaboration finally interact in a *business transaction*.

Linking between interacting transaction roles

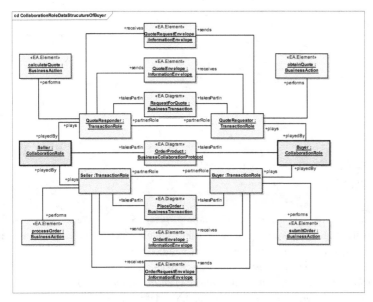

Fig. 9–3 The collaboration role data structure in regard to our order from quote example collaboration (conceptual)

By initiating the transformation process with the *order from quote* collaboration, the algorithm outputs the following object structure (depicted in Figure 9–3): two *collaboration role* objects are instantiated - one representing the *buyer* and one representing the *seller* (depicted with grey background). Both roles participate in a *business collaboration use case* called *order from quote* (Figure 9–2). Following the relationship between a *business collaboration use case* and a *business collaboration protocol*, we associate both *collaboration roles* with the *business collaboration protocol* describing *order from quote*. With respect to the *mapsTo* relationships shown in Figure 9–2 we create two *transaction role* objects for the *buyer collaboration role*: one *transaction role* for the *quote requestor* participating in *request for quote* and one *transaction role* for the *buyer* role participating in the *place order* transaction. In regard to the *seller collaboration role* we instantiate again two *transaction role* objects: one for the *quote responder* taking part in the *request for quote* transaction and one for the *seller* participating in the *place order* transaction. In addition, the algorithm associates each of the *transaction roles* with the respective *business trans-*

action, the *sent* and *received information envelope* and the *business action (requesting* or *responding business activity)* it performs as shown in Figure 9–3.

9.2.3 Creating WSDL descriptions

We create the WSDL instances capturing a partner's web service operations. In UMM *business transactions*, *requesting* and *responding business activities* depict the *operations* a role consumes and provides. Each *business action* (either a *requesting* or *responding business activity*) results in a WSDL *operation*. *Input* and potential *output* of a WSDL *operation* are determined by the flow of *information envelopes* between the two *business actions*. Operations provided by the same *collaboration role* in multiple *business transactions* are merged into one *port type*. Thus, each *collaboration role* results in exactly one *port type*. In our *order from quote* example we have a *buyer port type* and a *seller port type* conducting a collaborative order process. The interfaces of both *collaboration roles* - *buyer* (Listing 9–1) and *seller* (Listing 9–2) - are described in their own WSDL instances, containing the assembled *port type* and the exchanged messages.

```
[320]  <wsdl:message name="PurchaseOrderResponseEnvelope" />
[321]  <wsdl:message name="BusinessSignalAckReceipt" />
[322]  <wsdl:message name="BusinessSignalAckProcessing" />
[323]  <wsdl:message name="BusinessSignalControlFailure" />
[324]  <wsdl:portType name="BuyerPortType">
[325]    <wsdl:operation name="ReceiveResponseFor_submitOrder">
[326]      <wsdl:input message="tns:PurchaseOrderResponseEnvelope" />
[327]    </wsdl:operation>
[328]    <wsdl:operation name="AckReceipt">
[329]      <wsdl:input message="tns:BusinessSignalAckReceipt" />
[330]    </wsdl:operation>
[331]    <wsdl:operation name="AckProcessing">
[332]      <wsdl:input message="tns:BusinessSignalAckProcessing" />
[333]    </wsdl:operation>
[334]    <wsdl:operation name="ControlFailure">
[335]      <wsdl:input message="tns:BusinessSignalControlFailure" />
[336]    </wsdl:operation>
[337]  </wsdl:portType>
[338] </wsdl:definitions>
```

Listing 9–1 Order from quote example: The port type of the buyer and the exchanged messages

Generating WSDL descriptions follows the approach defined by [HH04]. Since participants in a transaction interact via their *business actions* (either a *requesting* or a *responding business activity*), each *business action* provides a corresponding service. It follows, that each *business action* results in its own *operation* within the respective partner's *port type*.

In case of a *requesting business activity* we are compliant to [HH04] and add a *ReceiveResponse* to the *operation* name. If a *business transaction* is synchronously executed, only the *responding business activity* results in an *operation*. In this case, the *operation* specifies an *output message* in addition to the *input message*. Finally, since a *business transaction* might require the transmission of business signals, we add *operations* to communicate an *acknowledgement of receipt*, an *acknowledgement of processing* and a *control failure* signal to each *port type*.

[339]	`<wsdl:message name="PurchaseOrderEnvelope" />`	
[340]	`<wsdl:message name="QuoteRequestEnvelope" />`	
[341]	`<wsdl:message name="QuoteEnvelope" />`	
[342]	`<wsdl:message name="BusinessSignalAckReceipt" />`	
[343]	`<wsdl:message name="BusinessSignalAckProcessing" />`	
[344]	`<wsdl:message name="BusinessSignalControlFailure" />`	
[345]	`<wsdl:portType name="SellerPortType">`	
[346]	`  <wsdl:operation name="processOrder">`	
[347]	`    <wsdl:input message="tns:PurchaseOrderEnvelope" />`	
[348]	`  </wsdl:operation>`	
[349]	`  <wsdl:operation name="calculateQuote">`	
[350]	`    <wsdl:input message="tns:QuoteRequestEnvelope" />`	
[351]	`    <wsdl:output message="tns:QuoteEnvelope" />`	
[352]	`  </wsdl:operation>`	
[353]	`  <wsdl:operation name="AckReceipt">`	
[354]	`    <wsdl:input message="tns:BusinessSignalAckReceipt" />`	
[355]	`  </wsdl:operation>`	
[356]	`  <wsdl:operation name="AckProcessing">`	
[357]	`    <wsdl:input message="tns:BusinessSignalAckProcessing" />`	
[358]	`  </wsdl:operation>`	
[359]	`  <wsdl:operation name="ControlFailure">`	
[360]	`    <wsdl:input message="tns:BusinessSignalControlFailure" />`	
[361]	`  </wsdl:operation>`	
[362]	`</wsdl:portType>`	

Listing 9–2 Order from quote example: The port type of the seller and the required messages

As introduced before, we use the WSDL data binding features provided by the *.NET framework* in our implementation. The namespace *System.Web.Services.Description* contains the required classes to construct a WSDL instance. The *service description* class corresponds to the root element (*definitions*) of a WSDL description. It defines references to every other object representing a WSDL element.

In the *order from quote* example we create two *service description*s - one for the *buyer* collaboration role and one for the *seller* role. Next, we set the WSDL namespace to the corresponding namespace of the particular *collaboration role* defined by its *namespace URI* property. Next step is to add a *port type* object to each service description. The *port type* object comprises the *operation* objects.

Creating service descriptions for the order from quote example

Firstly, we have a look at the *request for quote* transaction, which follows the *query/response* pattern. The *query/response* pattern indicates that the responder has the requested information available prior to the request. Hence, the response is transmitted synchronously back, because no further processing is needed. In regard to the *seller's port type* the synchronous execution results in an *operation* object called *calculate quote* (line 349). *Calculate quote* takes a *quote request envelope* as input (line 350) and outputs a *quote envelope* (line 351). *Calculate quote* requires *message* objects - one named *quote request envelope* (line 340) and one named *quote envelope* (line 341). As we outlined before, synchronous transactions result only in an *operation* on the responder's side. Thus, we add nothing for the *request for quote* transaction to the *buyer's port type*.

Processing request for quote

The second *business transaction - place order* - is specifed as *commercial transaction*. This mandates that the transaction is executed asynchronously. Furthermore the partners exchange business signals in order to acknowledge receipt and processing of received information. Concerning the *seller's* side (Listing 9–2), its *responding business activity* results in an *operation* object called *process order* (line 346). *Process order* expects a *purchase order envelope* as input. This results in a *message purchase order envelope* (line 339), which is referenced as *input* message by the *process order operation* (line 347). In regard to the other role, the *buyer* provides an *operation* to receive the response for the purchase order. According to the convention introduced above, we name the operation object *receive response for submit order* (line 325). *Receive response for submit order* takes a *purchase order response envelope* as input. Thus, we add a *purchase order response envelope message* object (line 320) to the *buyer's service description* and refer to it from the corresponding *operation* (line 326). Moreover, according to the business transaction pattern both parties provide *operations* to get an *acknowledgement of receipt* (lines 328 and 353) and an *acknowledgement of acceptance* (lines 331 and 353) for the envelopes they send. An additional *operation* needs to be provided by both parties to receive failure notifications (lines 334 and 359). The respective business signals are again specified by corresponding *message* objects (lines 321 to 323 and 342 to 344) and added to both *port types*.

Deriving a WSDL description for the place order transaction

As outlined above, a *business transaction* might be executed synchronous or asynchronous. Since UMM defines currently no service binding layer, we make some assumptions to fill this gap in the mapping process. A *business transaction* might be synchronous if the information is available prior to the exchange and creating the response requires no extensive further processing. Following this definition, the *request/confirm* and the *query/response* pattern are qualified to be processed synchronously. In our UMM Add-In implementation the modeler has two choices: *business trans-*

Synchronous and asynchronous execution of business transactions

actions might be transformed following the approach described above or asynchronously independent of their pattern.

9.2.4   Generating partner link types

The next transformation step creates *partner link types* for each collaboration role. BPEL uses the concept of *partner link types* to specify a binary
communication relationship between two services. The two interacting
services are described by their *roles* and their *port types*. The role of service
is determined by the context of the business collaboration (e.g. buyer, seller,
financial institution,...). The *port type* contains the *operations* that are relevant to the respective binary interaction. If a BPEL process specifies a multiparty collaboration, one *partner link type* is defined for each binary relationship within the multiparty collaboration.

Partner link types specify conversational relation-ships between two services

Partner link type definitions might either be placed within the WSDL
description or within a seperate artifact. The extensibility mechanism of
WSDL 1.1 allows us to define a *partner link type* as a direct child of the
WSDL root element (*definitions*). Defining *partner link types* within the
WSDL artifact enables reusing the WSDL *target namespace* definition.

For constructing *partner link types* with the UMM Add-In we utilize
again the *collaboration role data structure*. We identify binary conversational relationships by iterating over *transaction roles*. Each unique combination between two *collaboration roles* results in its own *partner link type*.

Our *order from quote* example collaboration is composed of two *business transactions*. Since it is a binary collaboration, there is only one unique
combination of *collaboration roles*. Hence, the *order from quote* collaboration results in exactly one *partner link type* named *buyer seller link type*
(Listing 9–3).

```
[363] <PartnerLinkType name="BuyerSellerLinkType">
[364]   <role name="Buyer">
[365]     <portType name="BuyerPortType" />
[366]   </role>
[367]   <role name="Seller">
[368]     <portType name="SellerPortType" />
[369]   </role>
[370] </PartnerLinkType>
```

Listing 9–3 Order from quote example: Partner link type describing the conversational relationship between buyer and seller

The roles - *buyer* and *seller* - are specified by the *name* attribute in the
corresponding *role* elements. Within each *role* element the role's *port type* is
referenced.

The UMM Add-In implementation puts *partner link types* directly into
WSDL artifacts. The *.NET framework* allows the definition of additional

WSDL elements via a class called *extension*. The *partner link type* is encapsulated within an *extension* instance and appended to the *service description* of both partners. Moreover, in order to facilitate access to the *partner link type* in the following steps, we add the *partner link type* object to our *collaboration role data structure* as well.

9.2.5   Generating BPEL process descriptions

BPEL describes a business process from a particular partner's point of view. In contrast, UMM specifies a collaborative process from a common point of view neutral to any participant. Hence, we need to derive a BPEL description for each participant in the collaborative process. The resulting BPEL processes must be complementary to each other.

A process description that is derived via our UMM to BPEL transformation engine follows the structure given in Listing 9–4. A BPEL process description starts with the definition of relationships between business partners by means of *partner links* (Line 372 to 374). Subsequently, it captures *variables* (Line 375 to 377) that are used within the process. Variables are utilized to reference documents, which further point to document structures in a WSDL, or for managing retries in case of message exchange failures. The actual choreography of a *business collaboration protocol* is defined within a *flow* container (Line 378 to 386). *Business transactions* are specified in *sequences* (Line 383 to 385) within the *flow* container. Each *business transaction* is thereby described in its own scope of a *sequence* activity. BPEL *links* (Line 380 to 382) choreograph the execution order of the transactions within the *flow*. Moreover, additional constructs like *switches*, *forks* and *joins* might be introduced within a *flow* in order to map more complex choreographies.

```
[371] <bpws:process>
[372]   <bpws:partnerLinks>
[373]     <!-- Partner links defining interactions with partners -->
[374]   </bpws:partnerLinks>
[375]   <bpws:variables>
[376]     <!-- Variables; mainly used for exchanged documents -->
[377]   </bpws:variables>
[378]   <bpws:flow>
[379]     <!-- a business collaboration protocol is mapped to a flow -->
[380]     <bpws:links>
[381]       <!-- links map to transitions of a business collaboration protocol -->
[382]     </bpws:links>
[383]     <bpws:sequence>
[384]       <!-- A sequence maps to a business transaction -->
[385]     </bpws:sequence>
```

Listing 9–4 Structure of a BPEL description as generated by the UMM Add-In

```
[386]   </bpws:flow>
[387] </bpws:process>
```

Following the algorithm of our implementation, in the next section we describe relationships between the process participants via *partner links*. Then we focus on the choreography defined by a *business collaboration protocol* and identify the *flow* of *business transactions*. The most complex step - the transformation of *business transactions* - follows afterwards.

Specifying partner links

Partner links specify the conversational relationship between the process owner and the partners he collaborates with. *Partner links* are comparable to instances of *partner link types*. A *partner link type* determines which role is played by the process owner and which role is taken up by a partner. Similar to a *partner link type*, a *partner link* refers to exactly two *roles*. A *partner link* references the *roles* defined within the *partner link type* and assigns them to the participating partners.

The UMM Add-In implementation identifies *partner links* by means of the *collaboration role data structure*. The algorithm iterates over each *collaboration role* object and retrieves its set of *partner link types*. One *partner link* object is created for each *partner link type*. Naming of the newly created *partner links* follows the convention „*LinkTo{OtherRole}*", whereby *{OtherRole}* is replaced by the name of the partner role. The attribute *my role* refers to the process owner's role and the *partner role* attribute is set to the name of the collaborating partner role. In addition, the *partner link* refers to the corresponding *partner link type* via the attribute *partner link type*.

Generating partner links in the transformation engine

Considering our *order from quote* example we gather two *partner links*. Listing 9–5 shows the *partner link* for the *seller's* process, whereas the *partner link* within the *buyer's* process is given in Listing 9–6. According to the convention introduced before, the *seller's partner link* is named *link to buyer* (Line 389). Line 390 references the corresponding *partner link type*. Since the *seller* is the process owner the *myRole* attribute (Line 391) refers to his role. It follows, that the buying role is performed by its partner (Line 391).

```
[388] <bpws:partnerLinks>
[389]     <bpws:partnerLink name="LinkToBuyer"
[390]         partnerLinkType="SellerBuyerLinkType"
[391]         myRole="Seller" partnerRole="Buyer" />
[392] </bpws:partnerLinks>
```

Listing 9–5 Order from quote example: partner link for the seller's process

In the *buyer's* BPEL process the *partner link* is named *link to seller* (Line 394). Since the two *partner link* definitions must be compliant, the *partner link type* attribute refers to the same *seller buyer link type* (Line 395). Now the *buyer* owns the process, hence the roles specified by the *myRole* and the *partnerRole* attribute are permuted (Line 396).

```
[393] <bpws:partnerLinks>
[394]     <bpws:partnerLinkname="LinkToSeller"
[395]         partnerLinkType="SellerBuyerLinkType"
[396]         myRole="Buyer" partnerRole="Seller" />
[397] </bpws:partnerLinks>
```

Listing 9–6 Order from quote example: partner link for the buyer's process

Transforming transitions into links

BPEL provides the concept of *links* to choreograph *activities* within a *flow*. A *link* connects exactly two *activities*. A condition may guard the execution of a *link*. As soon as the condition evaluates true, the *link* is activated. *Activities* that are target of multiple *links* might define *join conditions*. Such a *join condition* has to evaluate true in order to execute the *activity*. If no *join condition* is defined, there is an implicit *join condition* applied that requires at least one *link* to be positive.

The choreography of a *business collaboration protocol* is mapped to a BPEL *flow* construct. The execution order of a *business collaboration protocol* is specified by means of UML *transitions*. In BPEL, *links* choreograph a *flow* of *activities*. It follows, that UML *transitions* within a *business collaboration protocol* map to *links* in a BPEL *flow*.

Links in a BPEL process correspond to UML transitions

In our implementation, deriving *links* starts with querying over all *collaboration roles*. For each *collaboration role* we retrieve the *business collaboration protocol* the role takes part in. By iterating over all *transitions* that are part of a *business collaboration protocol* we determine candidates for BPEL *links*. Since BPEL defines no constructs for capturing start and end states of a process we omit *transitions* that are either connected to *initial* or *final states* in a *business collaboration protocol*. Hence, only *transitions* that are connected on both ends to either *activities*, *decisions*, *forks* or *joins* are transferred into *links*.

Describing the implementation

After its creation, a BPEL *link* is additionally wrapped into an object structure (called *link wrapper)* capturing the name of both its *source* and *target activity*. We do this due to the fact, that a *link* is just a semantic construct with an arbitrary name. An *activity* includes a child element *source* for each *link* that starts from this *activity*. Similarly, an *activity* includes a child element *target* for each *link* that leads to this *activity*. The *link name* attribute of a *source* or *target* element refers to the corresponding *link*.

A *link* itself provides no possibility to retrieve its *source* or *target activity*. However, in further processing steps we need to identify a *link's source* and *target activity* for defining a choreography.

After relevant *transitions* are transformed into BPEL *links* we add them to the process specification as depicted in Listing 9–4. Furthermore, we persist the *link wrapper* objects in the scope of the particular *collaboration role*.

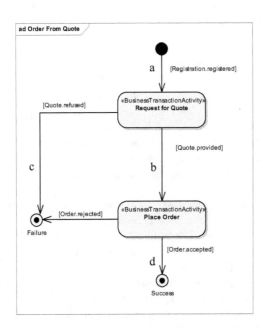

Fig. 9–4 Example collaboration: Only transition b results in a BPEL link

The *business collaboration protocol order from quote* (Figure 9–4) contains five UML *transitions*. However, all *transitions (a, c, d, e)* except one *(b)* - which leads from *request for quote* to *place order* - are connected to an *initial* or *final state*. Hence, we discard them. The remaining *transition (b)* is transformed to a corresponding BPEL *link*. As the example in line 398 shows, our implementation follows the convention *{SourceName}_to_{TargetName}* for naming *links* between *activities* in a BPEL process.

[398] <bpws:link name="RequestforQuote_to_PlaceOrder" />

Transforming business transactions in general

Deriving BPEL descriptions of *business transactions* is the most complex stage within the transformation engine. In BPEL, a UMM *business transaction* is represented by a *sequence* built of a set of *basic activities* denoting the information exchange.

Invoke activities are utilized to signalize a service invocation. *Invoke* activities correspond to sending an *information envelope*. Receiving a document from a partner via one's own web service interface is specified by a *receive* activity. In case of a synchronous interaction, we depict a partner's response via a *reply* activity. By their nature *reply* activities must always follow *receive* activities.

In order to group *activities* that are performed within the scope of a *business transaction* we use the concept of a *sequence* in BPEL. A *sequence* mandates that contained *activities* are executed serially. A *business transaction* is specified by either a one-way or two-way information exchange which might additionally include the transmission of business signals. Business documents (i.e. *information envelopes*) and possible business signals are exchanged in a predefined and sequential execution order. Thus, the concept of a *business transaction* matches the requirements of a BPEL *sequence*. Converting a *business transaction* to BPEL requires the transformation of exchanged *information envelopes* - the actual business information - as well as business signals. Each business signal results in its own *message* as well as in its own *operation*.

The scope of a business transaction is mapped to a sequence. Information exchange is depicted via invoke, receive and reply activities

Within a *sequence* we specify preceding and succeeding *activities* by referring to *links*. *Links* have already been identified in stage *Transforming transitions into links*. Now we profit from the *link wrapper* objects, which persist the *link* itself as well as its *source* and *target activity*. By looping over the set of *link wrappers* we compare the *business transaction* to the *source* and *target activity* specifications of each *link wrapper*. If the transaction equals the source of the *link* we create a corresponding *source* object with reference to the *link*. If the transaction is target to the *link*, we add a corresponding *target* object that refers to the *link*.

Considering *order from quote*, the algorithm results in two *sequence* stubs as shown in Listing 9–7. Both *business transactions - request for quote* and *place order* - result in their own *sequence* (lines 399 to 403 and 404 to 407). The *sequence* name is specified according to the one of the corresponding transaction. Further we specify the incoming and outgoing connections of both *sequences*. *Request for quote* refers to the link *request for quote to place order* via a *source* element (line 400). *Place order* references to the same *link* via a *target* element (line 405). It follows, that *request for quote* directly precedes *place order* in the execution order. In addition, a *transition condition* is assigned to the *link* (line 401) between the two

Determining the choreography of the order from quote example collaboration

sequences. The condition postulates that a *quote is provided* in order to evaluate true. Hence, a quote must be issued prior to an order submission.

```
[399] <bpws:sequence name="RequestforQuote">
[400]     <bpws:source linkName="RequestforQuote_to_PlaceOrder"
[401]             transitionCondition="Quote.provided" />
[402]                 <!-- activities specifying message exchange -->
[403] </bpws:sequence>
[404] <bpws:sequence name="PlaceOrder">
[405]     <bpws:target linkName="RequestforQuote_to_PlaceOrder" />
[406]                 <!-- activities specifying message exchange -->
[407] </bpws:sequence>
```

Listing 9–7 Sequence stubs representing the two business transactions of the order from quote example collaboration

So far, we generated *sequence* stubs representing the scope of a *business transaction* as well as the relationships between *business transactions* in the context of a choreography. Furthermore, we have outlined which basic BPEL *activities* might be utilized to map the choreography of a *business transaction.*

The following two subchapters discuss how a *business transaction* is described in the language of BPEL. We start discussing the implementation by means of the requesting side and afterwards we look at the responding side.

As other stages in the BPEL generation, transforming *business transactions* is tightly bound to roles. By iterating over *collaboration roles* and their roles in business transactions, the algorithm identifies and transforms *business transactions* that are of relevance to a particular collaboration role. Since BPEL describes a process from a partner's point of view, the code for the requestor differs from the one of the responder.

The generated code depends if a role is the initiator or the responder of a business transaction

Transforming the requestor's part of a business transaction

The requesting role of a *business transaction* initiates the information exchange. In case of a failure the requestor usually reinitiates the information exchange. The number of retries is specified by the *retry count* of the *requesting business activity.* If the *retry count* equals zero - this means no retries are required - the algorithm continues with generating the *invoke* statement that transmits the request envelope.

If a *retry count* greater than zero is specified we create a BPEL structure that allows the retransmission of the *requesting information envelope.* We basically create a loop that decrements a variable holding the available retries each time a failure occurs. If the retries are exhausted, we exit the loop. The *variable* that represents the *retry count* is created outside of the *sequence* at the same location where the document *variables* are specified (line 375 to 377 in Listing 9–4). In the *sequence* the *variable* is initialized

Generating code that allows to reinitiate a business transaction

with the corresponding *retry count* using an *assign* activity. The *assign* activity allows to copy an expression - i.e. the number of retries - to the *retry count variable*. Next, the algorithm creates a loop using a *while* activity. The execution of the *while* loop is guarded by a condition. The *condition expression* is specified by the *retry count variable* and the *variable representing the received business document*. The condition evaluates true, if the *retry count variable* has a value greater than zero and if the *variable representing the last expected response document* is not null. It is null until the document is received. The last expected document might either be the *responding information envelope* in case of a two-way transaction or an *acknowledgement* in case of an one-way transaction. In case of a timeout exception within the *business transaction*, the retry count is decremented. We leave the *while* loop, if the *retry count variable* holds zero or if the last expected document has been received. The transaction is considered to be successful if the document is received and unsuccessful if no retry is left. In order to determine success, the algorithm checks after the *while* loop if the *retry count* equals zero. The check is implemented via a nested *case* clause in a *switch* activity. The query of the *retry count variable* is assigned to the *case condition*. If the condition evaluates true - i.e. there are no more available retries - a fault is actuated via a *throw* statement.

```
[408] <bpws:assign>
[409]     <bpws:copy>
[410]         <bpws:from expression="3" />
[411]         <bpws:to variable="processOrderRetryCount" />
[412]     </bpws:copy>
[413] </bpws:assign>
[414] <bpws:while name="CheckprocessOrderRetries"
[415]         condition="getVariableData('processOrderRetryCount') &gt; 0 AND
[416]         getVariableData('PurchaseOrderResponseEnvelope')= NULL">
[417]     <bpws:sequence>
[418]      <bpws:invoke partnerLink="LinkToSeller" portType="SellerPortType"
[419]      operation="processOrder" inputVariable="PurchaseOrderEnvelope" />
[420]     <bpws:pick>
[421]         <bpws:onMessage partnerLink="LinkToSeller"
[422]             portType="BuyerPortType" operation="AckReceipt">
[423]             <bpws:empty />
[424]         </bpws:onMessage>
[425]         <bpws:onAlarm for="PT2H">
[426]             <!-- decrement process order retry count -->
[427]         </bpws:onAlarm>
[428]     </bpws:pick>
[429]     <bpws:pick>
[430]         <bpws:onMessage partnerLink="LinkToSeller"
[431]             portType="BuyerPortType" operation="AckProcessing">
[432]             <bpws:empty />
```

Listing 9–8 BPEL snippet which describes the buyer's part of the place order transaction

```
[433]              </bpws:onMessage>
[434]              <bpws:onAlarm for="PT6H">
[435]                   <!-- decrement process order retry count -->
[436]              </bpws:onAlarm>
[437]         </bpws:pick>
[438]         <bpws:pick>
[439]              <bpws:onMessage partnerLink="LinkToSeller"
[440]                   portType="BuyerPortType"
[441]                   operation="ReceiveResponseForSubmitOrder"
[442]                   variable="PurchaseOrderResponseEnvelope">
[443]                   </bpws:empty>
[444]              </bpws:onMessage>
[445]              <bpws:onAlarm for="PT24H">
[446]                   <!-- decrement process order retry count -->
[447]              </bpws:onAlarm>
[448]         </bpws:pick>
[449]         <bpws:sequence>
[450]              <bpws:invoke partnerLink="LinkToSeller"
[451]                   portType="SellerPortType"
[452]                   operation="AckReceipt" />
[453]              <bpws:invoke partnerLink="LinkToSeller"
[454]                   portType="SellerPortType"
[455]                   operation="AckProcessing" />
[456]         </bpws:sequence>
[457]         </bpws:sequence>
[458] </bpws:while>
[459] <bpws:switch>
[460]         <bpws:case
[461]              condition="bpws:getVariableData('processOrderRetryCount') == 0">
[462]         <bpws:throw faultName="bpws:processOrderControlFailure" />
[463]         </bpws:case>
[464] </bpws:switch>
```

Listing 9–8 shows the BPEL code describing the *buyer's* part of the *place order* transaction. In order to create a complete BPEL specification of the *order from quote* collaboration, the code in line 408 to 464 replaces line 406 in Listing 9–4. In line 408 to 413 we assign a value of three to the corresponding *retry count variable*. The *while* loop spanning from line 414 to 458 contains the *activities* describing the buyer's part of a transaction. The *while* loop continues as long as the *retry count variable* holds a value greater than zero and no *purchase order response envelope* has been received. In case of a time out exception during the information exchange the *retry count variable* is decremented. When we exit the loop we check the value of the *retry count variable* using a *switch* activity (lines 459 to 464). If the condition (line 461) equals true, a fault is raised (line 462).

The requestor of a transaction initiates the interaction by invoking the appropriate partner's service. This is denoted by an *invoke* activity. The *invoke* activity requires the name of the *operation*, the partner's *port type* that contains the operation description, the *partner link*, and *possible input* and *output variables*. The *requesting information envelope* sent to the responder's service is denoted using the *input variable*. In case of a synchronously executed two-way transaction the *information envelope* sent back to the requestor is specified by the *output variable*. No *output variable* is needed in case of an asynchronous two-way transaction or a one-way transaction.

Using invoke to send the requesting information envelope

Depending on the business transaction pattern the requestor might require *acknowledgements of receipt* and *processing*. If the transaction follows a two-way pattern the requestor expects a *response information envelope*. For each type of response - no matter whether it is a business signal or a business document - we add a corresponding *pick* structure after the *invoke* statement. A *pick* activity allows the definition of multiple events. Only the first occurring event and its associated activities are executed. Within the *pick* activity we use *onMessage* elements to specify messages the *buyer* might be receiving in this stage. Via the attributes of *onMessage*, the requestor's own *port type* is specified in conjunction with the *operation* that is called by the responder. If the *onMessage* element specifies the receipt of a *responding information envelope* (i.e business information) we additionally define a *variable* to persist the message. Otherwise, if a business signal is received we omit persisting it and hence no *variable* is specified. Moreover, the algorithm specifies an *onAlarm* event within the *pick* activity. After a given time period, an *onAlarm* event executes if no *onMessage* event happened. We specify the duration to wait for business signals or *information envelopes* using the *onAlarm's for* attribute. Within the *onAlarm* element, the algorithm creates an *assign* activity in order to decrement the *retry count variable*. We add a *pick* construct for each message that might be received according to the business transaction flow. It follows, that we add at most three *pick* elements - one for the *acknowledgement of receipt*, one for the *acknowledgement of processing* and one for the *response information envelope* - in a sequential order.

Processing the response using pick activities

Considering the *place order* transaction (Figure 9–5 and Listing 9–8) from the *buyer's* point of view the information exchange takes place as follows:

Fig. 9–5 Order from quote example: business transaction place order

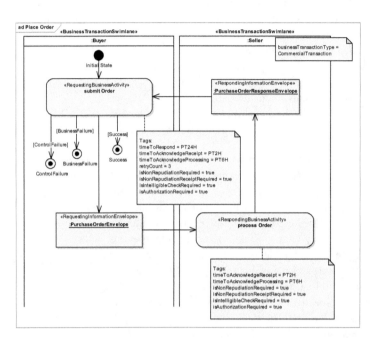

According to the choreography of the *business transaction* the *buyer* invokes the *seller's process order* operation (lines 418 and 419). The *operation* as well as the *seller's port type* are referenced within the *invoke* activity. Furthermore, a *purchase order envelope* is required as input to the *process order operation* (line 419). After the *purchase order envelope* is sent to the *seller's* service, the *buyer* expects an *acknowledgement of receipt*. This results in a *pick* activity (line 420 to 428) with a nested *onMessage* (line 421 to 424) element. *OnMessage* refers to an *operation* of the *buyer's port type* (line 422), because the *buyer* waits for an incoming call of its service. If a corresponding acknowledgement is received we actually do nothing - denoted by the *empty* activity in line 423. This ends the *pick* activity and we proceed to the next *pick*. However, if the *buyer* receives no *acknowledgement of receipt* within a duration of 2 hours, the *onAlarm* construct executes (line 425 to 427). Within the *onAlarm* construct (line 435) we decrement the *retry count variable*.

If an *acknowledgement of receipt* has been properly picked up, the *buyer* claims an *acknowledgement of processing*. We denote waiting for an *acknowledgement of processing* (line 429 to 437) similarly to that of the *acknowledgement of receipt*. However, the *seller* has now a timeframe of 6 hours (line 434) in order to respond with an appropriate acknowledgement.

Describing place order from the buyer's point of view in BPEL

After the *seller* has processed the *purchase order envelope* he responds with a *purchase order response envelope*. Since the receipt of the response message is also time-critical we utilize again a *pick* activity (line 438 to 448).

The structure of the *pick* activity for response messages is again similar to that for picking up acknowledgements. However, the *purchase order response envelope* must be persisted in a corresponding *variable*. The *buyer* waits 24 hours for the *purchase order response envelope* (line 445) as specified by the *time to respond* of the *requesting business activity*. In case no *purchase order response envelope* is communicated by the *seller* within the 24 hours timeframe, the *retry count variable* is again decremented (line 446).

As specified by the *responding business activity*, the *seller* requires the *buyer* to acknowledge receipt of the *purchase order response envelope*. The *buyer* acknowledges the receipt of the *purchase order response envelope* via the *invoke* activity in line 450 to 452. Afterwards, when the *purchase order response envelope* is handed over to the business application, the buyer communicates an *acknowledgement of processing* (line 453 to 455). Since the *buyer* transmits business signals to the *seller*, the *buyer* invokes *operations* provided by the *seller's* service. Thus, the *seller's port type* and its *operations* to receive business signals are referenced within both *invoke* activities.

The code in Listing 9–9 describes the *buyer's* process with respect to *request for quote* (Figure 9–6). Listing 9–9 replaces line 402 of Listing 9–4. *Request for quote* requires the *buyer* to reinitiate the transaction in case of timeout exceptions. Hence, we use again a *variable* to represent the *retry count*. *Obtain quote* specifies a *retry count* of three, which is assigned to a corresponding *variable* (lines 465 to 470). In addition, we use a *while* loop (line 471 to 479) and a *switch* activity (line 480 to 485) similar as for *place order* to check the yet available retries .

Waiting for the response information envelope is depicted by a pick activity

```
[465] <bpws:assign>
[466]       <bpws:copy>
[467]             <bpws:from expression="3" />
[468]             <bpws:to variable="calculateQuoteRetryCount" />
[469]       </bpws:copy>
[470] </bpws:assign>
[471] <bpws:while name="CheckcalculateQuoteRetries"
[472]       condition="bpws:getVariableData('calculateQuoteRetryCount') &gt; 0
[473]       AND bpws:getVariableData('QuoteEnvelope') == NULL">
[474]       <bpws:sequence>
[475]       <bpws:invoke partnerLink="LinkToSeller" portType="SellerPortType"
[476]             operation="calculateQuote" inputVariable="QuoteRequestEnvelope"
[477]             outputVariable="QuoteEnvelope" />
```

Listing 9–9 Order from quote example: code snippet describing request for quote from the buyer's point of view

```
[478]      </bpws:sequence>
[479] </bpws:while>
[480] <bpws:switch>
[481]      <bpws:case
[482]         condition="bpws:getVariableData('calculateQuoteRetryCount') == 0">
[483]             <bpws:throw faultName="bpws:calculateQuoteControlFailure" />
[484]      </bpws:case>
[485] </bpws:switch>
```

However, sending and receiving messages in *request for quote* differs
from *place order* in two ways. At first, the invocation of the *seller's calculate quote* operation is executed synchronously. This is signalized by the
presence of the *output variable* attribute (line 477). The attribute value
denotes that the response of the synchronous interaction is assigned to a
variable named *quote envelope*.

No *pick* and nested *onMessage* elements are required as for receiving an
asynchronous response. Furthermore, *request for quote* requires no transmission of business signals. Neither the *buyer* nor the *seller* requires an
acknowledgement of receipt or *processing* from the other partner. Thus, we
specify no *picks* to wait for acknowledgements and no *invoke* statements for
calling the *seller's* business signal operations.

Transforming the responder's part of a business transaction

The semantics of a *business transaction* requires that a responder's first task
is receiving an envelope from the initiator. In BPEL, we denote the receipt
of a message by a *receive* activity. The service interface that receives the
message is specified via the *operation* and *portType* attribute. The *variable*
attribute is again used to persist the incoming message. The *partnerLink*
attribute refers the respective conversational relationship.

If the responder is required to acknowledge receipt or processing of a
requesting information envelope, our transformation engine adds the corresponding *invoke* activities.

In case of a two-way business transaction pattern the responder transmits business information back to the requestor. In case of a synchronous
response, we use a *reply* activity. A *reply* activity references the same *port
type* and the same *operation* as its preceding *receive* activity. Hence, both
activities refer to the responder's *port type* and to the *operation* that receives
the *requesting information envelope*. If the transaction is executed asynchronously an *invoke* activity is used to send the response message. The
invoke activity refers to the requestor's *port type* and to the corresponding
operation provided by the *requesting business activity* in order to receive
the response.

*The responding information
envelope might be transmitted synchronously or asynchronously*

If the responder requires business signals from the requestor, we create *pick* structures that allows the process to wait for an acknowledgement. If a certain duration is exceeded, a timeout exception is thrown. The structure and semantics of such *pick* structures are introduced in detail in *Transforming the requestor's part of a business transaction*. However, the semantics of a *business transaction* defines no *retry count* on the responders side. If no acknowledgement is received in the scheduled duration, the responder resends the message. Resending the message is carried out until the proper acknowledgement is received or the transaction's time limit is exceeded.

```
[486] <bpws:receive partnerLink="LinkToBuyer" portType="SellerPortType"
[487]      operation="processOrder" variable="PurchaseOrderEnvelope" />
[488] <bpws:sequence>
[489]      <bpws:invoke partnerLink="LinkToBuyer"
[490]          portType="BuyerPortType" operation="AckReceipt" />
[491]      <bpws:invoke partnerLink="LinkToBuyer"
[492]          portType="BuyerPortType" operation="AckProcessing" />
[493] </bpws:sequence>
[494] <bpws:invoke partnerLink="LinkToBuyer"
[495]      portType="BuyerPortType"
[496]      operation="ReceiveResponseForSubmitOrder"
[497]      inputVariable="PurchaseOrderResponseEnvelope" />
[498] <bpws:pick>
[499]      <bpws:onMessage partnerLink="LinkToBuyer"
[500]          portType="SellerPortType" operation="AckReceipt">
[501]              <bpws:empty />
[502]      </bpws:onMessage>
[503]      <bpws:onAlarm for="PT2H">
[504]              <!-- re-send message -->
[505]      </bpws:onAlarm>
[506] </bpws:pick>
[507] <bpws:pick>
[508]      <bpws:onMessage partnerLink="LinkToBuyer"
[509]          portType="SellerPortType" operation="AckProcessing">
[510]              <bpws:empty />
[511]      </bpws:onMessage>
[512]      <bpws:onAlarm for="PT6H">
[513]              <!-- re-send message -->
[514]      </bpws:onAlarm>
[515] </bpws:pick>
```

Listing 9–10 Order from quote example: BPEL snippet describing place order for the seller

Listing 9–10 shows the *seller's* process of the *place order* transaction (see Figure 9–5). The *seller* receives a *purchase order envelope* via an operation called *process order* (line 486 and 487). The seller confirms the receipt by sending an *acknowledgement of receipt* (line 489 to 490). In addition, when the message is handed over to the business application he communicates an *acknowledgement of processing* (line 491 to 492). When the purchase order is processed accordingly, the seller sends a *purchase order response envelope* back to the *buyer's* service (line 494 to 497). The *seller* requires the *buyer* again to acknowledge receipt and processing of the *purchase order response envelope*. The corresponding *picks* are denoted in line 498 to 515).

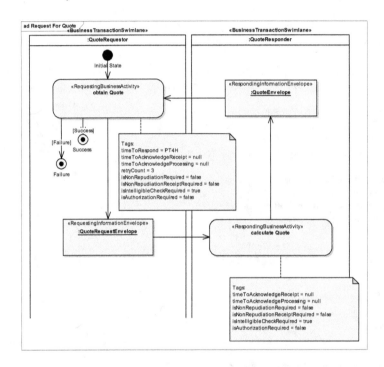

Fig. 9–6 Order from quote: request for quote transaction

[516] <bpws:receive partnerLink="LinkToBuyer" portType="SellerPortType"
[517]   operation="calculateQuote" variable="QuoteRequestEnvelope" />
[518] <bpws:reply partnerLink="LinkToBuyer" portType="SellerPortType"
[519]   operation="calculateQuote" variable="QuoteEnvelope" />

Listing 9–11 Order from quote example: BPEL snippet describing the seller's request for quote activities

Considering *request for quote* (Figure) from the *seller's* point of view (Listing 9–11), we add only two *activities*. The *seller* receives the *quote request envelope* via the *calculate quote* operation (line 516 to 517). Since *request for quote* is executed synchronously, the *quote envelope* is sent back via a *reply* activity (line 518 to 519). As outlined before, since a *reply* activity denotes a response to an operation call, it must refer to the same *operation* and *port type* as the preceding *receive* activity.

9.3   Conclusion and outlook

In this chapter we introduced a technical implementation to transform UMM collaborations to BPEL process descriptions. We discussed conceptual mapping rules as well as implementation details specific to our engine.

The current version of our transformation engine outputs only *abstract processes*, which are also called *business protocols* in BPEL. Hence, we create process description on a rather conceptual level. Since UMM includes currently no service binding definitions, implementation layer details with respect to protocol bindings and service endpoints are currently not considered. A service binding layer for UMM will provide the missing link creating executable BPEL processes directly from UMM *business collaboration models*. As soon as such a specialization module is defined for the UMM foundation, we will incorporate it into the UMM Add-In.

10   Mapping Business Information to Document Formats

In chapter 2 we have already seen, that UMM is a process centered approach. Nevertheless UMM also endeavors to consider the data which is exchanged during the business process. In this chapter we will see which different approaches are possible in order to model the data which is exchanged between the business partners. An introduction into the concepts of Core Components (CCTS) and Universal Business Language (UBL) will be given. Furthermore an example will be presented, where an XML schema is generated which serves as a specification for the data to be exchanged during the process.

Modeling and mapping business data

For the time being the mapping of business information to document formats is still in an early alpha stage of development. Therefore a major part of this chapter is a theoretical approach which has not been implemented yet. However current effort is invested in the completion of the business information mapping feature and future releases of the UMM Add-In will contain the full functionality.

10.1   Introduction to business information modeling

Process oriented standards like ebXML and UMM offer methods to describe the profiles, processes and procedures of business processes. However what is often missing is a commonly accepted standard to describe the business information which is exchanged during a business process. What one cannot find in the standards mentioned above is for example a specification of an invoice or a purchase order.

However in the past UN/EDIFACT has presumably been the most important standard in the field of business information. In recent years other standards have evolved as well and proven to be a serious competitor to UN/EDIFACT. Although thoroughly developed and maintained over the years UN/EDIFACT has some shortcomings which new exchange standards try to overcome. The following paragraphs will introduce two well known standards namely the Universal Business Language (UBL) which is

New competition for the old bull UN/EDIFACT

based on XML and the Core Components Technical Specification (CCTS). Furthermore the particular advantages of the XML based standards will be shown. CCTS will be the first standard to be explained because UBL is build upon CCTS.

10.2   Core Components Technical Specification (CCTS)

In this subchapter we would like to give a short introduction to the concept of Core Components and how they interrelate with the concept of *business information entities* which are used to model information within UN/CEFACT's Modeling Methodology. CCTS itself forms the basis for another well known standard namely the Universal Business Language (UBL) which we will treat in the next chapter. As this chapter is supposed to be an introduction and does only give a superficial insight into the CCTS standard we would like to redirect the interested reader to the Core Components Technical Specification [CC03] for further and more detailed information about Core Components.

Applications in the field of e-business are often lacking an interoperability in regard to the information which is exchanged. Prior approaches of information standardization were often focused on the static definition of messages. What is needed is a new approach which allows a flexible and interoperable way of standardizing business semantics. *Considering the business semantics*

The CCTS describes a way to identify, capture and maximize the way of business information reuse in a syntax neutral manner. It will ensure that two trading partners which are using different syntaxes (e.g. UN/EDIFACT and XML) use business semantics in the same way. This requires, that both partners use a syntax which is based on the same Core Components. Hence a clear mapping between different messages is feasible with no regard to syntaxes, regional or industry boundaries.

As a central building block for the standard serves the concept of a Core Component. A Core Component is a building block for an information exchange package which contains all information pieces that are necessary to describe a specific concept.

We distinguish between four different types of *core components*: *Different core components*

- Basic Core Component (BCC)
- Core Component Type (CCT)
- Aggregate Core Component (ACC)
- Association Core Component (ASCC)

A *basic core component* represents a single characteristic of a specific *aggregate core component*. Its business semantic definition is unique. In *Basic core component*

Figure 10–1 the *aggregate core component* Person.Details has two *basic core components*, namely Name and Text. These *basic core components* are the so called properties of the *aggregate core component*. Therefore in this context a *basic core component* is often also referred to as a *basic core component property*. However both definitions imply the same. A *basic core component property* has an associated *data type*. The *basic core component* Name is of type *text* and the *basic core component* Birth Date is of type *date*.

One can also regard a *basic core component* as an attribute of an *aggregate core component*. Those familiar with the object oriented approach will find the comparison with a class helpful. An *aggregate core component* can be regarded as a class and its *basic core component properties* as the attributes of the class. However please note, that the object oriented concept has nothing to do with the CCTS specification. The comparison is only supposed to encourage the principles of CCTS.

In Figure 10–1 Name and Birth Date are the *basic core components* of the *aggregate core component* Person.Details.

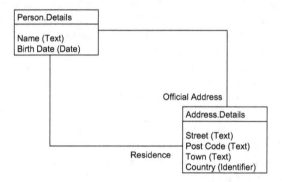

Fig. 10–1 Core component example

A collection of related pieces of business information which together form a distinct business meaning is called an *aggregate core component*. In Figure 10–1 Person.Details and Address.Details are *aggregate core components*.

Aggregate core component

A *core component* which represents a complex characteristic of an *aggregate core component* is called an *association core component*. It has a business semantic meaning which is unique. The *association core component* is represented by a property and associated to an *aggregate core component* which describes its structure. In Figure 10–1 we can see two *association core components* namely Residence and Official Address. Both are properties of the *aggregate core component* Person.Details and both are associated with the *aggregate core component* Address.Details.

Association core component

A *core component* containing only one *content component* which carries the content plus optional *supplementary components* is called a *core component type*. *Supplementary components* give an extra definition to the *content component*. A *core component type* does not have any business semantics.

Core component type

The *core component Type* `Amount.Type` for instance has a *content component* which carries the value `12`. The value per se has no meaning on its own. By adding a *supplementary component* with the value `Euro` we are giving a meaning to the *content component*.

The properties of *core components* are defined by *data types* which represent a full range of values that can be used for the representation of a particular property. A *data type* must be based on a *core component type*. The *data type* is defined by specifying restrictions which limit the use of the *core component type's* values which is the base of the *data type*.

Data types

As we saw until now, *core components* define an abstract concept which helps to model business information. By introducing the business context, we can qualify and refine *core components* according to their use within particular business circumstances.

Introducing the business context

The Core Components Technical Specification covers two significant areas, the *core components* and the *business information entities*, which are depicted in Figure 10–2.

Fig. 10–2 Overview about the core components Technical Specification

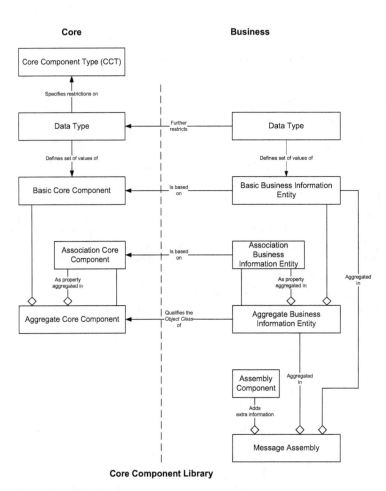

Core Component Library

As we can see, a *basic core component* becomes the basis of a *basic business information entity* when it is used in a real business circumstance. Therefore a *business information entity* is the outcome of a *core component* being used in a specific business context. The same holds for the *association core component* and for the *aggregate core component. data types* which are used in the business context are *data types* taken from the *core components* and restricted to the needs of the business modeler. As already mentioned above the *data type* used for the *core components* is a restricted *core component type.*

Basic business information entities and *aggregate business information entities* are aggregated in a *message assembly.* In order to add additional information to the stored *information entities, assembly components* are used.

As with *core components*, we can also distinguish different *business information entities*:

*Different business informa-
tion entities*

Basic Business Information Entity (BBIE)
Association Business Information Entity (ASBIE)
Aggregate Business Information Entity (ABIE)

A *basic business information entity* represents a *basic business information entity* property and is linked to a specific *data type*. A *basic core component* is the basis of a *basic business information entity*. One can regard a *basic business information entity* as a property of an *aggregate business information entity*. In Figure 10–2 Name and Birth Date are *basic business information entity*. They are properties of the *aggregate business information entity* US_Person.Details.

*Basis business information
entity*

*Fig. 10–3 Business
information entity
example*

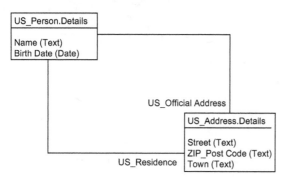

A collection of business information pieces which are related and together form a business meaning in a specific business context is called an *aggregate business information entity*. In Figure 10–3 US_Address.Details and US_Person.Details are *aggregate business information entities*.

*Aggregate business infor-
mation entity*

A *business information entity* which represents a complex characteristic of an *aggregate business information entity* is called an *association business information entity*. It has a business semantic meaning which is unique. The *association business information entity* is represented by a property and associated with an *aggregate business information entity* which describes its structure. In Figure 10–3 the two attributes US_Official Address and US_Residence are *association business information entities*. They are properties of the *aggregate business information entity* US_Person.Details.

*Association business infor-
mation entity*

The concept of CCTS as it has been described above is an abstract one which cannot be used in practice directly. An concrete implementation of the standard has to be found in order to integrate the concept into a business data modeling process. In the next chapter we will introduce the concept of the Universal Business Language which builds on the standards defined by the Core Components Technical Specification. The upcoming chapters will then focus on naming an design criteria and on a specific reference implementation for Enterprise Architect.

10.3   Universal Business Language (UBL)

UBL was found by the Organisation for the Advancement of Structured Information Standards (OASIS) with the aim to define business information exchanged in an arbitrary business process. During the year 1999 efforts were made within OASIS in order to create a set of standard XML documents. At the end of the year UN/CEFACT and OASIS began their collaboration on ebXML. The ebXML standard was released without specifying, in which format the business data has to be exchanged between the two business partners. Hence the working group which came to be known as UBL started its work, focusing on an XML based business document standard.

A first implementation of the abstract CCTS concept

UBL 1.0 was released as an OASIS standard and is available free of charge from the organization's homepage. Just like ebXML the newly introduced standard is based on the XML Schema Definition Language [XSD04]. The core of the standard is a set of XML schemes for the description of business documents. Within UBL no messaging standard for the exchange of the business information is defined. It is up to the implementor to choose an exchange form which could for instance be ebXML, SMTP or SOAP.

The UBL standard is divided into various categories containing reoccurring components which facilitate their reuse. In addition eight document types are predefined. The eight document types are:

- Despatch Advice
- Invoice
- Order
- Order Acknowledgement
- Order Cancellation
- Order Response (Complex)
- Order Response (Simple)
- Receipt Advice

Furthermore code lists e.g. for currency abbreviations or country codes are included. In UBL recurring text components are named *basic information entities* (BIEs). They consist of atomic (*common base components*) and aggregated components (*common aggregate components*).

The specification of UBL very much considers efforts already made in the field of EDI. From the technical point of view the *basic information entities* are based on the Core Components Technical Specification, which we examined in the last chapter. However CCTS is an abstract concept which is not implemented in a specific language like UBL.

Furthermore UBL is build on the xCBL - XML Common Business Library [xCB03]. The main aim of xCBL was the representation of UN/EDIFACT messages with an XML vocabulary. The xCBL initiative has announced that with UBL 1.0 being released, the xCBL standard will merge with UBL.

Reasons for taking xCBL as the basis for UBL

OASIS itself mentions several reasons, why the standard was initially build on xCBL and not developed from scratch. First of all, xCBL already was a widespread and commonly accepted standard which was in use in a number of enterprises. SAP and CommerceOne for instance were already using xCBL. Furthermore xCBL was based on a library concept which guaranteed a better alignment of the document types derived from the library in contrary to a standard, where the documents are developed indenpendently. A third argument was the fact, that xCBL was developed on a kind of open source basis, which allowed the easy extension of the standard. As we already saw in the last chapter, the CCTS standard only provides an abstract definition. UBL is a true implementation of the standards defined by CCTS. The UBL library uses the *business information entities* which are defined in CCTS. Current effort is taken by the UBL working group in order to map UBL to the standard Core Component Library. Furthermore research is undertaken together with UN/CEFACT Applied Technology Group (ATG) and Open Applications Group (OAGI) in order to develop a single XML standard for the representation of core component types and unqualified data types.

Relationship of UBL to CCTS

UN/CEFACT and OASIS already collaborated during the creation of the ebXML standard. It has been stated by UN/CEFACT that

Relationship of UBL to UN/CEFACT

> "UN/CEFACT will support only one document-centric approach to XML content, and its desire is that UBL will be the foundation for that approach"

Current negotiations are also about the transfer of UBL from OASIS to UN/CEFACT.

Nevertheless one could now scrutinize, where the specific advantage of UBL is compared to practiced and well known standards like UN/EDIFACT. Whereas UN/EDIFACT is mainly used by large companies which can afford parsers and software specifically designed for the use of UN/EDIFACT, small and medium sized enterprises are not having such resources at their disposal. UN/EDIFACT parsers are difficult and complex

UBL vs. UN/EDIFACT

to implement whereas a XML based format can be easily managed using DOM-parsers available in almost every programing language. In addition all along industry specific dialects of UN/EDIFACT have evolved which are not compatible to each other e.g. VDA [Ver] and ODETTE [Ode]. Hence the necessity of a new standard which is eligible for an industry no matter what size gets apparent.

Another very important argument is the need for an extension mecha-   *Extension made easy*
nism. XSD offers extensibility through its inheritance mechanism. It allows to define specific components based on basic components. Country and industry specific changes can therefore be easily inferred from standardized base types.

If for example a japanese company delivering supply for a US IT-industry wants to extend the UBL basic component `<address>` by an additional attribute without obstructing the collaboration process with the US company the process would be the following:

> create a new type `<addressJapanese>` which is inferred from the basic type
>
> in all new documents used by the japanese company a reference to he newly created type `<addressJapanese>` is made

The software of the US company will still operate properly because it is working with the generic address type. The additional attribute is only used by the japanese company. Therefore the US company does not need to implement any changes in their IT systems. However if the US company wants to access the newly created attribute of the japanese company changes in the software of the US company are required as well. Such an extension mechanism would not be possible in such an easy manner if the two companies were using UN/EDIFACT.

In UN/EDIFACT one can distinguish between two cases which might occur during an extension. In the first case, the additional field which the trading partners want to use is already part of the standard. In this case the message implementation guide has to be modified accordingly. In the second case, the additional field is not part of the standard. In this case a request to change the standard has to be initiated. It will last at least 6 months until the change in the UN/EDIFACT standard is proceeded.

Another important factor is the fact, that XML uses Unicode messages. Hence the depiction of other languages like asian ones is feasible as well.

As mentioned before, UBL wants to eliminate the boundaries between   *UBL as a chance for small*
the enterprises' ability to perform B2B messaging. Especially for small and   *and medium sized enter-*
medium sized companies an appropriate software in order to participate in a   *prises*
B2B process is the first threshold. One of the most diffused commercial of the shelf software (COTS) is Microsoft Office 2003 which contains an

XML compliant tool named Infopath. Figure 10–4 depicts a typical B2B document exchange between two companies.

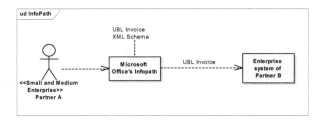

Fig. 10–4 UBL eliminates
B2B boundaries

On the left hand side we can see the small enterprise of partner A, who can not afford an enterprise system and is therefore using Microsoft Infopath in order to submit an UBL Invoice to the large enterprise of partner B. Partner A uses Microsoft Infopath and the UBL invoice XML schema and creates a form. Infopath is capable of storing the created XML document but it can also submit it using e-mail or a web service. The ERP system of partner B can then automatically process the invoice of partner A. This is only one of numerous examples, how UBL could be used between two business exchanging information.

In the next chapter we will focus on the need of naming and design rules. Without applying strict rules to business documents an uncontrolled growth of custom naming and design patterns would occur.

10.4   The need for Naming and Design Rules (NDR)

In order to guarantee a standardized naming and design convention for XML documents, UN/CEFACT Applied Technology Group (ATG) has specified a set of naming and design rules. These rules serve as an allegation for XML documents generated from business information modeled in Enterprise Architect. In this chapter we would like to stress the significance of such rules and explain the most important ones in detail. For the interested reader we would like to refer to the XML Naming and Design Rules [ATG05] published by ATG2.

Before we start to immerse into the naming and design rules we are going to examine the relationship between *core components, business information entities* and *XSD artifacts*. The naming and design rules are closely coupled to the concept of *core components* and *business information entities*. Figure 10–5 shows the relationship between the concept of *core components, business information entities* and *XSD artifacts* at a glance. The relationship between the context neutral and the context specific part has already been thoroughly explained in the chapter about Core Components.

Naming and design rules are essential for business information mapping

We will therefore focus on the relationship of the syntax specific part to the context neutral and context specific part.

Fig. 10–5 The relation between CCs, BIEs and XSD

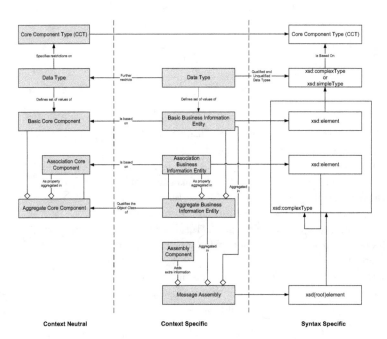

The *message assembly* in which the *aggregate business information entities* are aggregated is realized as a xsd:complexType definition and global element in the UN/CEFACT XSD Schema. The type of the global element declaration is based on the xsd:complexType which represents the document level *aggregate business information entity*. The global element is the root element of any XML instances conformant to the XSD Schema.

An *aggregate business information entity* is defined as a xsd:complexType as well. We already saw in the chapter about Core Components, that *association business information entities* are references to other *aggregate business information entities*. Furthermore an *association business information entity* is an attribute of an *aggregate business information entity*. Within an XSD Schema this concept is realized by a local element which is declared within a xsd:complexType. The local element represents the associating *aggregate business information entity*. Hence the content model of the associated *aggregate business information entity* is included in the content model of the associating *aggregate business information entity*.

Like an *association business information entity* a *basic business information entity* is defined as a local element within a `xsd:complexType`. The type of the `xsd:complexType` refers to the *aggregate business information entity*, which contains the *basic business information entity*.

Data types can be defined as `xsd:complexType` or as `xsd:simpleType`. As described in the upper right corner of Figure 10–5 *data types* are based on *core component type* `xsd:complexType` from the CCT schema module.

The *core component types* are aggregated in the *core component type* schema module. This schema module is the normative XSD expression of CCTS *core component type*.

Concerning the naming rules the XML standard implicitly imposes rules, which naturally apply to any XML document. As an example Listing 10–1 shows a sample XML file. Although it might appear valid for those not so familiar with XML it apparently is not. An XML element name must not begin with a numeric character.

```
[520] <?xml version="1.0" encoding="UTF-8"?>
[521] <schema>
[522]      <23></23>
[523] </schema>
```

Listing 10–1 A sample invalid XML structure

This is just one example for a rule which is already specified by the XML standard itself. For those not so familiar with the XML standard we would like to refer to [XML04]. The naming and design rules specified by ATG2 are build upon the XML standard by applying restrictions on it.

For the naming of elements and attributes, the ATG2 rules follow the ebXML Architecture Specification's best practice. For attributes lower camel case (LCC) is used and for elements upper camel case (UCC) is used. Listing 10–2 gives an example for both design requirements.

Naming elements and attributes

```
[524] <xsd:attribute name="unitCode"/>
[525] <xsd:element name="LanguageCode"/>
```

Listing 10–2 Example for LCC and UCC

Line 524 shows an example for an attribute which is named after the lower camel case convention. In line 525 we can see an example for an element which is named after the upper camel case convention. It is a common practice especially in the field of programming to write compound nouns joined without spaces and each word capitalized. The name 'camel' comes from the association of the alternating up and down of the capital letters with a camel's back.

In order to allow a reusability of the schema components a modular model has been chosen by ATG. Modules are choosen because they are

Ensuring the reusability

unique in their functionality or because they represent a part of a bigger schema in order to enhance manageability and performance. Figure 10–6 shows an overview of UN/CEFACT's XSD Schema Modularity Scheme.

Fig. 10–6 UN/CEFACT's XSD Schema Modularity Scheme

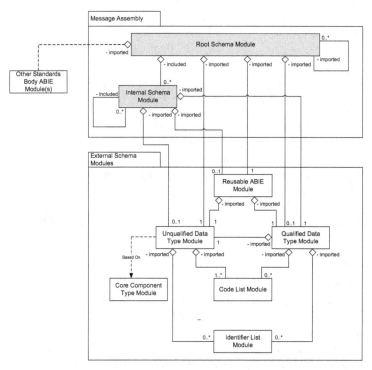

The scheme is divided into two significant parts namely the *message assembly* and the *external schema modules*. Within the *message assembly* the *root schema module* is defined. A *root schema module* can include several *internal schema modules* which have to be in the same namespace as the *root schema*. Furthermore the *root schema module* itself can import other *root schemas* from different namespaces. Other standards body ABIE modules can be imported into the *root schema module* as well.

Message Assembly

The *external schema modules* on the other hand are from another namespace than the *root schema*. The *external schema module* consists of a set of reusable *ABIE modules, qualified data types, unqualified data types* and modules containing *code lists* and *identifier lists*. Please note, that each module within the *external schema module* resides in its own namespace. The *root schema module* always imports the *qualified data type module*, the *unqualified data type module* and the *reusable ABIE module*. The other associations within Figure 10–6 are to be read accordingly.

External schema modules

Another important issue concerning the creation of XML documents is the usage of namespaces. Namespaces provide a simple method which allows qualifying elements and attributes by associating them with namespaces. Namespaces are identified by URI references and enable the interoperability and consistency of XML artifacts for the use in a library of reusable schema modules and types. As already proved in the last paragraph, the namespace concept is essential in order to guarantee the correct relationship between the different schema modules. Figure 10–7 gives an overview about the namespace scheme as it is used by UN/CEFACT. The scheme provides a flexible and robust approach which allows the definition of UN/CEFACT specific namespaces.

The significance of XML namespaces

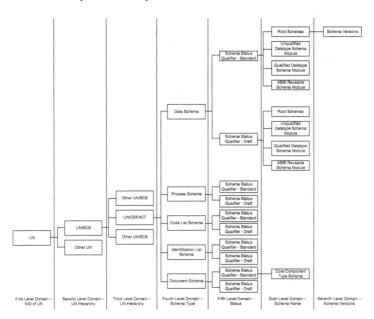

Fig. 10–7 UN/CEFACT's namespace scheme

Listing 10–3 shows an example namespace URI derived from the schema in the figure above. Line 526 shows the building pattern for a namespace. Line 527 gives an example for a UN/CEFACT namespace. The schematype in line 527 is *data* and the name of the module is *Modulexyz*. The major version is *0*, the minor version *3* and its revision is *6*.

[526] urn:un:unece:uncefact:<schematype>:draft:<name>:<major>.[<minor>].[<revision>]
[527] urn:un:unece:uncefact:data:draft:Modulexyz:0.3.6

Listing 10–3 Sample namespace scheme

With the namespace description the short introduction into the XML naming and design rules finishes. The next chapter will show a reference implementation within the UMM Add-In which merges the concepts of CCTS, UBL and the XML naming and design rules.

10.5 A reference implementation

Normally the last artifact which is finished by the modeler is the *business information view*, which represents the information that is exchanged during the business process. UMM itself does only define a very loose structure for the information itself and leaves it up to the modeler, which standard is implemented. Figure 10–8 shows the business information structure as it is provided by the UMM meta model. As we can see, the *information entity* serves as the superstructure for all business information exchanged during a business process. The information which is included in an *information envelope* is composed of exactly one *information entity* which serves as a header. Furthermore there can be one or more *information entities*, which serve as the body of the message. *information entities* can be nested recusively. Please note, that the *information envelope* is an *information entity* as well.

Business information and UMM

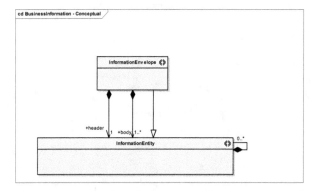

Fig. 10–8 The business information view at a glance

We will now examine, how business information can be modeled by using UN/CEFACT's Core Components Technical Specification. Furthermore we will focus on how the CCTS specification is integrated into Enterprise Architect.

10.5.1 CCTS Profile

As we saw in chapter 10.2, CCTS is a concept on an abstract basis. It is up to the user how he implements the CCTS standard. An implementation can

Using CCTS within Enterprise Architect

be done by using spread sheets or regular documents. Nevertheless in order to efficiently implement the CCTS standard into the UMM Add-In a method must be found which allows an integration into Enterprise Architect. Enterprise Architect itself admits the definition of UML Profiles. Such profiles allow to tailor the modeling language to specific areas - in this case to Core Components.

A major issue is the integration of UN/CEFACT's Core Component Library (CCL) into enterprise architect. Because the Core Component Library will be stored in an UN/CEFACT repository and identified by an ebXML compliant registry current efforts focus on a registry binding as well. For the time being the standard procedure is importing the Core Component Library which is stored in an XMI format into Enterprise Architect. The according *aggregate core components* and *data types* are then available via the tree view in Enterprise Architect. Although convenient in regard to complexity, the disadvantage of a Core Component Library in the form of a simple import into Enterprise Architect gets apparent. Changes in the Core Component Library are not incorporated in Enterprise Architect directly. An architecture with a service binding to the library which is stored in a repository and identified by a registry would guarantee up to date *core components*. Figure 10–9 shows the idea of a registry binding. For the sake of lucidity the repository has been omitted.

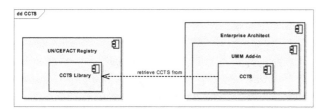

Fig. 10–9 A registry binding for core components

The workflow for the modeler would be the following. First the modeler creates the process model with the according views. As a last artifact the user creates the business information view which holds the information to be exchanged during the business process. Before the user starts to model the information he retrieves the current *core components* from the CCTS library. Enterprise Architect connects to the registry and retrieves the information about the *core components*. Within Enterprise Architect the information from the registry is transformed into modeling elements which allows utilizing the CCTS for modeling. Hence the modeler always has *core components* which are up to date. The whole process could be implemented automatically as well. If for instance the user starts Enterprise Architect or marks a specific model as UMM model, the tool could automatically load the CCTS information required from the registry.

That of course takes a permanent internet connection as granted, which however can be anticipated nowadays.

10.5.2 CCTS modeling in practice

After the CCTS profile has successfully been imported into Enterprise Architect, the modeler is given every feature needed in order to depict business information in Enterprise Architect. The import of the CCTS profile is done automatically when the user marks the model as UMM model. The workflow which is described in the following paragraph currently has to be done manually. However a programmatic support by the UMM Add-In is in development.

During the modeling procedure, the user drag and drops the specific *aggregate core components* from the Core Components Library into the model and specifies their parameters. By implementing the *aggregate core component* into a business context, the *aggregate core component* becomes an *aggregate business information entity*.

Modeling the business information

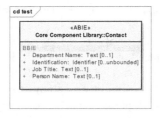

Fig. 10–10 Aggregate core component

Figure 10–10 shows a sample *aggregate core component* with the according *basic core components*. In the next step the modeler deletes the *basic business information entities* of the *aggregate business information entities*, which are not needed. Then the modeler adds qualifier terms to both the *aggregate* and the *basic core components* to create *aggregate business information entities* and *basic business information entities*.

Fig. 10–11 Aggregate business information entity

Figure 10–11 shows the *aggregate business information entity* which has been derived from the *aggregate core component* shown in Figure 10–10.

At this step it can happen, that some *basic core components* will be completely removed from the corresponding *aggregate business information entity* whereas other could become multiple *basic business information entities*. If the latter occurs, each must have a different qualifier term.

In the next step the modeler decides, whether he needs any *qualified data types* (QDT) or not. In order to create a *qualified data type* the modeler restricts a *core data type* (CDT) to his needs.

Following the data definition, the modeler updates each *basic business information entity* as necessary which means he changes *core data types* to *qualified data types*.

The final step towards a business information model is done when the modeler aggregates the *aggregate business information entities* into an assembly document. Given the finished business information model, the user is ready to transform the information model into a data structure of choice which will presumably be an XML schema. The next paragraph will focus on the issue of transforming a business information model into an XML schema representation.

10.5.3   CCTS validation and transformation

As already shown in the chapter about the validator, a valid model is a prerequisite for a transformation. Just like a UMM model which has to be valid in order to generate a BPSS or a BPEL out of it, a business information model must be valid as well, before a transformation procedure can be invoked.

Figure 10–12 shows an example for a *business information view*, which has been modeled with UBL *business information entities*. Before we can start to transform the *business information view*, we have to make sure, that the overall structure of the view is correct. The transformation engine anticipates a valid business information model and would throw an exception if an invalid model is passed to it.

Fig. 10–12 A business
information example
provided by Red Wahoo

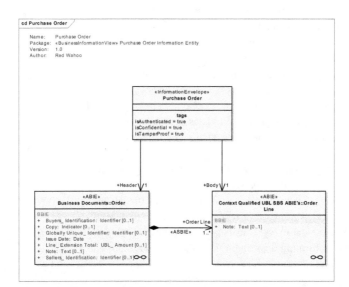

The user clicks on the *business information view* which he wants to tran-
form. The validator then first checks the *information envelope* and the corresponding header and body information. As one can see, the header and
body of the *information envelope* are nested elements. Hence the algorithm
iterates through the graph recursively. Any errors which occur during the
validation are presented to the user in the well known interface which we
already saw in chapter 8. If no error occurs, a new window pops up, which
allows the modeler to specify additional information for the transformation
procedure. Figure 10–13 shows the business information transformer interface at a glance.

Fig. 10–13 The business
information transformer
interface

In the upper left corner the selected *business information view* is shown. Underneath the user can specify the root element for the generated XSD schema. In the settings box, the user has the chance to apply additional preferences to the generated schema e.g. which name and design rules the generated schema must follow. By clicking *Generate Schema* the generation process is invoked and any messages are presented to the user in the status box. If no error occurred during the transformation the user is presented a dialog window where he must choose where to save the generated schema.

The only setting which is currently supported is the generation of a schema which follows the naming and design rules specified by ATG2. Future releases will include more settings.

10.5.4   CCTS import feature

In general it can be said, that Enterprise Architect is an efficient tool for modeling business processes. However in order to model business information it is lacking some features which are better implemented in other tools. These features are for example the advanced derivation of *business information entities* from *core components* as well as their storage and retrieval.

Getting the business information in Enterprise Architect

A lot of third party tools which are currently on the market allow the modeling of EDI specific information. In order to reach interoperability with these tools, current effort is spend on an import interface. The idea is, that information which has been modeled in a third party tool can be imported into Enterprise Architect and used in order to model the business information. The reference implementation which is currently in an early alpha phase targets at the integration of data exported from EDIFIX into Enterprise Architect. Figure 10–14 shows the deployment diagram of a reference implementation.

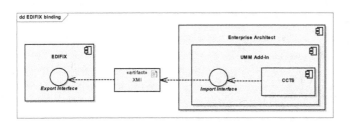

Fig. 10–14 An Enterprise Architect - EDIFIX binding

The business data modeler creates *core component* data within EDIFIX and then exports the data into an XMI format using the export interface of EDI-FIX. In the next step the user imports the XMI based EDIFIX data into Enterprise Architect using the import interface of the UMM Add-In. During the import process the *core components* which have been modeled within EDIFIX are transformed into an Enterprise Architect modeling elements. Hence the user can use the freshly imported *core components* to model the business information needed.

Another attempt focuses on the export of *core component* data from Enterprise Architect into EDIFIX. However current research has shown, that this approach might not be feasible as the information which is provided by Enterprise Architect is less than the one required by EDIFIX.

Data export feature

As we have seen, the modeling of business information plays a very important role in the business process modeling. Without the specification of an exchange format which is agreed upon by the business partners an information exchange cannot take place. The UMM Add-In focuses on the support of the modeler in his ability to model business information. Furthermore the mapping of business information into exchange formats is pursued.

11 Summary and Outlook

In this thesis a thorough description of UN/CEFACT's Modeling Methodology (UMM) was given. We started with identifying problems and needs in real-world B2B scenarios. Based on these needs we depicted the requirements for a methodology like UMM in order to denote cross-enterprise business processes. Participating in such an interorganizational process requires a partner to agree on a certain choreography. Furthermore, a participant must provide compliant interfaces to his information systems according to the agreed choreography. However, if each participant describes the same process just from its own view, the resulting process descriptions would not match. Thus, we require an approach - like UMM - in order to specify a process from a global point of view.

Moreover, we depicted the need for an adequate tool support in order to endorse modelers in producing UMM compliant models. Afterwards, we presented some technical details in respect to our implementation called UMM Add-In. The UMM Add-In is an extension to the UML tool Enterprise Architect written in .NET.

After giving some general information about UMM's history and the responsible standardization body - the United Nations Centre for Trade Facilitation and Electronic Business (UN/CEFACT) - we introduced worksheets and their role in UMM to capture business knowledge. In addition, we described the integration of a worksheet editor into the UMM Add-In as well as the benefits thereof.

The next chapter comprises a guide supporting the creation of a UMM compliant model based on the know-how gathered by worksheets. The guide explains step by step the modeling of each view in the UMM. Furthermore, the required steps to draw up a UMM model are illustrated using an example about an ordering process.

However, giving only a guide for modeling UMM might result in models, which are not compliant to the UMM meta model defined by the UML Profile for UMM. This is not the modeler's failure, but a presumable result when applying a formal notation like UMM without a computational verification. Hence, we illustrated the need for validating UMM models and gave an introduction into the UML Profile for UMM. Furthermore, we presented the implementation of a validation engine as part of the UMM Add-In based on the constraints defined in the UML Profile for UMM.

In a service-oriented architecture, XML-based process descriptions are utilized to configure information systems according to a particular process. In our case, we have already modeled collaborative processes by means of UMM. It follows, that an automatic generation of process descriptions is required in order to support the deployment of B2B information systems. In the UMM Add-In we implemented a transformation engine for generating BPEL compliant artifacts. BPEL seems to be the winner amongst the set of lately emerged choreography languages.

Modeling exchanged business information and deriving document schemes thereof is the last building block to gain a complete set of artifacts that together describe a collaborative process. Describing the derivation of schemes as well as a reference implementation thereof according to rules standardized by UN/CEFACT is subject to the following chapter.

Finally, in the appendix we included the UML Profile for UMM, denoting the UMM meta model. During the writings of our thesis we attended the UN/CEFACT conferences and were responsible for the development of this profile.

However, we think that with our thesis a cornerstone has been laid, which allows the development of additional features. This might include the development of a registry connector in order to support the registration and retrieval of business collaboration models or parts thereof.

We hope, that the UMM Add-In and this thesis will help to increase the diffusion of UMM within the business process modeling world. Furthermore, our work should be a step forward to achieve the transition from a data-centric business modeling to a process-centric modeling of interorganisational business processes.

I. Appendix - Business Transaction Patterns

UN/CEFACT mandates the use of one of the six business transaction patterns that are already standardized by *RosettaNet* [ROS02]. These business transaction patterns cover every real-world business case. The list below describes each pattern in detail:

Commercial transaction (two-way): Represents the typical „offer and acceptance" business interaction. A *commercial transaction* results in a residual obligation between two parties to fulfill the terms of a contract. In other words both parties enter into a commitment to fulfill their part of the contract. An example would be the submission of an order and receipt of a purchase order response.

The *commercial transaction* pattern constitutes that the responding party has to return an *acknowledgement of receipt* when receiving the *requesting information envelope*. The time frame within the *acknowledgement of receipt* has to be sent is specified by *time to acknowledge receipt* (of the *requesting business activity*). If the document passes a set of business rules and is handed over to the business application the responder has to send an *acknowledgement of processing*. The corresponding timeframe is specified by the *time to acknowledge processing* (of the *requesting business activity*). Furthermore, the responding party has to return the *responding information envelope* within the period defined by *time to respond* (of the *requesting business activity*). The requesting party has to re-initiate the *business transaction* in case the *time to acknowledge receipt, time to acknowledge processing* or *time to respond* is exceeded. The number of attempts is defined by the *retry count*. When the responding party answers with the *responding information envelope* the requestor has to issue an *acknowledgement of receipt* within the *time to acknowledge receipt* (specified in the *responding business activity*). If the *responding information envelope* passes again the business rules (e.g. grammar validation, sequence validation...) the requestor has to transmit an *acknowledgement of processing* to the responder. The allowed period is set by the *time to acknowledge processing* of the *responding business activity*. Both parties are required to authorize themselves (authorization is required by both business actions) and have to the sign their envelopes and business signals (as defined by *non repudiation required* and *non repudiation of receipt required* of both business actions).

Query/Response (two-way): This pattern describes the request of information that is available to the responder prior to the request. This might be a fixed data set inside a database or any kind of static information (e.g. a catalog).

The requestor initiates the transaction by submitting the request within a *requesting information envelope* to the responder. The responder has to provide the information within the period specified by *time to respond*. The requestor has to re-initiate the transaction as defined by the *retry count* if the responder is not answering within the given *time to respond*. No business signals and no non-repudiation requirements are necessary in the *query/response* pattern.

Request/Response (two-way): A transaction follows the *request/response* pattern if the requestor asks for information that requires some business processing on the responder's side. This includes information that needs to be dynamically assembled and hence cannot be returned immediately (i.e. non-static information). An example would be the request for a product quote. The *request/response* pattern results in no residual obligation between the two parties to fulfill the terms of a contract. Concerning the *request for quote* example, this inquiry leads to no commitment of the requestor to buy the quoted product. Similarly the responder does not pledge himself to have the quoted product available in case of a further order.

The *request/response* pattern specifies the exchange of a *requesting* and a *responding information envelope*. Non-Repudiation requirements as well as requiring business signals are optional, but not recommended using the *request/response* pattern. If either business signals or non-repudiation are required, they follow the same semantics as specified for the *commercial transaction* pattern.

Request/Confirm (two-way): This pattern should be used if the requesting partner asks for information that requires only confirmation in respect to previously agreed business contracts. An example might be the request of status information.

The requestor initiates the transaction by submitting the request document to confirm to the responder. Business signals or non-repudiation are not required by the responder. Anyway, the requestor re-initiates the transaction as defined by the *retry count* if the responder misses answering within the *time to respond*. Regarding the *responding information envelope*, the responder might require that the requestor sends an *acknowledgement of receipt* when he receives the confirmation response (within the timeframe specified by *time to acknowledge receipt*). Furthermore, the responder might require the requestor to authenticate himself and to guarantee the non-repudiation of the *acknowledgement of receipt*.

Information distribution (one-way): Represents an informal, unidirectional information transmission. An example would be an information about price discounts to customers.

Neither business signals nor non-repudiation or authorization requirements are allowed in the information distribution pattern. Since the receipt of the distributed information is not guaranteed no *retry count* must be claimed.

Notification (one-way): A formal, unidirectional sending of information. This pattern is applied if the requesting side has to inform the responding side about an irreversible business state. An example is the notification of a product shipment.

Since the notification transmittal is a formal action the requestor has to claim for an *acknowledgement of receipt* with the specified *time to acknowledge receipt*. Furthermore, the non-repudiation of a receipt is required. If the reacting party is not sending the business signal within the agreed *time to acknowledge receipt* the requesting party has to re-initiate the transaction as specified by the *retry count*.

Furthermore each business document has to be checked for readability by the receiver as defined by the value of *is intelligible check required* which is by default set to true for every document. Table shows the requirements on the responding party within the different transaction patterns. These requirements are specified in the *requesting business activity* (because the requestor demands the responder to fulfill these requirements). Similarly Table shows the requirements posed by the responding party to the requesting party. We specify them using the *tagged values* of the *responding business activity*, because the responder demands them to be fulfilled by the requestor.

Default assignment of tagged values for a requesting business activity

The following table (Table) shows the default assignment of *tagged values* for a *requesting business activity*. They denote the requirements on the responder in context of the six business transaction patterns.

Default tagged values for a requesting business activity

Requesting Business Activity	Time to Acknowledge Receipt	Time to Acknowledge Processing	Time to Respond	Authorization Required	Non repudiation required	Non repudiation of receipt	Retry count	is IntelligibleCheckRequired
Commercial Transaction	2hr	6hr	24hr	TRUE	TRUE	TRUE	3	TRUE
Request/Confirm	null	null	24hr	FALSE	FALSE	FALSE	3	TRUE
Request/Response	null	null	4hr	FALSE	FALSE	FALSE	3	TRUE
Query/Response	null	null	4hr	FALSE	FALSE	FALSE	3	TRUE
Notification	24hr	null	null	FALSE	FALSE	TRUE	3	TRUE
Information Distribution	null	null	null	FALSE	FALSE	FALSE	0	TRUE

Default assignment of tagged values for a responding business activity

The following table (Table) shows the default assignment of *tagged values* for a *responding business activity*. They denote the requirements on the requestor in context of the six business transaction patterns.

Default tagged values for a responding business activity

Responding Business Activity	Time Acknowledge Receipt	Time to acknowledge processing	Authorization Required	Non repudiation required	is IntelligibleCheck required	Non requpidiation of receipt
Commercial Transaction	2hr	6hr	TRUE	TRUE	TRUE	TRUE
Request/Confirm	2hr	null	TRUE	FALSE	TRUE	TRUE
Request/Response	null	null	FALSE	FALSE	TRUE	FALSE
Query/Response	null	null	FALSE	FALSE	TRUE	FALSE
Notification	null	null	FALSE	FALSE	TRUE	FALSE
Information Distribution	null	null	FALSE	FALSE	TRUE	FALSE

II. Appendix - Bibliography

[ATG05]   UN/CEFACT Applied Technology Group (ATG). *XML Naming and Design Rules*, February 2005. Draft 1.1a.

[BEA03]   BEA, IBM, Microsoft, SAP AG and Siebel Systems. *Business Process Execution Language for Web Services*, May 2003. Version 1.1.

[Bir05]   Birgit Hofreiter and Christian Huemer and Ja-Hee Kim. Choreography of ebXML Business Collaborations. *Information Systems and e-Business Management (ISeB)*, 2005.

[BML02]   Assaf Arkin. *Business Process Modeling Language*, November 2002. Version 1.0.

[BPS03]   UN/CEFACT TMG. *UN/CEFACT - ebXML Business Process Specification Schema*, 2003.

[BRJ04]   Grady Booch, James Rumbaugh, and Ivar Jacobson. *The Unified Modeling Language Reference Manual*. Addison-Wesley Professional, 2004.

[BRJ05]   Grady Booch, James Rumbaugh, and Ivar Jacobson. *Unified Modeling Language User Guide*. Addison-Wesley Professional, 2005.

[CBP03]   UN/CEFACT TBG 14. *UN/CEFACT - Common Business Process Catalog*, November 2003. Version 0.95 (Candidate for Version 2.0).

[CC03]   UN/CEFACT TMG. *Core Components Technical Specification - Part 8 of the ebXML Framework*, November 2003. v2.01.

[CI02]   World Wide Web Consortium (W3C). *Web Service Choreography Interface*, August 2002. Version 1.0.

[CL02]   World Wide Web Consortium (W3C). *Web Service Conversation Language*, March 2002. Version 1.0.

[CSO02]   UN/CEFACT. *Structure and Organization of the UN/CEFACT Permanent Working Group*, June 2002. TRADE/CEFACT/2002/8/Rev.1.

[EBB05]   OASIS. *ebXML Business Process Specification Schema TS*, 2005.

[ETA01]   OASIS, UN/CEFACT. *ebXML - Technical Architecture Specification*, February 2001. Version 1.4.

[FOU03]   UN/CEFACT TMG. *UN/CEFACT Modeling Methodology (UMM) Foundation Module Specification*, 2003. Candidate for 1.0, First Working Draft.

[HC93]   Michael Hammer and James Champy. Reengineering the corporation: A manifesto for business revolution. *Business Horizons*, 36(5):90–91, 1993.

[HH04]   Birgit Hofreiter and Christian Huemer. Transforming UMM Business Collaboration Models to BPEL. In *OTM Workshops*, pages 507–519, 2004.

[Hof05]   Birgit Hofreiter. *The Impact of Business Context on Business Collaboration Models, Choreography Languages, and Business Documents*. PhD thesis, University of Vienna, 2005.

[Jur04]   Matjaz Juric. *Business Process Execution Language for Web Services - BPEL4WS*. PACKT-Publishing, 2004.

[Kin04]   Ekkart Kindler. Using the Petri Net Markup Language for Exchanging Business Processes? Potential and Limitations. In Jan Mendling and Markus Nuettgens, editors, *XML4BPM 2004, Proceedings of the 1st GI Workshop XML4BPM - XML Interchange Formats for Business Process Management, Marburg (Germany)*, pages 43–60, March 2004.

[Mic06]   Michael Ilger and Marco Zapletal. An Implementation to Transform Business Collaboration Models to Executable Process Specifications. accepted at Multikonferenz Wirtschaftsinformatik 2006 (MKWI06), Passau, Germany; will be published in GI LNI, 2006.

[MN04]   Jan Mendling and Markus Nuettgens. Exchanging EPC
             Business Process Models with EPML. In Jan Mendling
             and Markus Nuettgens, editors, *XML4BPM 2004, Pro-*
             ceedings of the 1st GI Workshop XML4BPM - XML Inter-
             change Formats for Business Process Management at 7th
             GI Conference Modellierung 2004,Marburg (Germany),
             pages 61–80, March 2004.

[MO102]   OMG Object Management Group. *Meta-Object Facility*,
             2002. Version 1.4 Specification.

[MOF05]   OMG Object Management Group. *MOF 2.0/XMI Map-*
             ping Specification, 2005. Version 2.1.

[Naj02]   Farrukh Najmi. Web Content Management using the
             OASIS ebXML Registry Standard. Technical report, Sun
             Microsystems, February 2002.

[OCL03]   Object Management Group (OMG). *OCL 2.0 - OMG*
             Final Adopted Specification, October 2003.

[Ode]   Odette. *Achieving Supply Chain Excellence in the Auto-*
             motive Industry.

[OER95]   ISO. *Open-edi Reference Model*, 1995. ISO/IEC JTC
             1/SC30 ISO Standard 14662.

[OPE01]   ISO/IEC. *Business agreement semantic descriptive tech-*
             niques - Part 1: Operational aspects of Open-edi for
             implementation, August 2001. ISO/IEC FDIS 15944-1.

[Pel03]   Chris Peltz. Web Services Orchestration and Choreogra-
             phy. *IEEE Computer*, 28(10):46–52, 2003.

[ROS02]   RosettaNet. *RosettaNet Implementation Framework:*
             Core Specification, December 2002. V02.00.01.

[SBD04]   UN/CEFACT Applied Technologies Group. *Standard*
             Business Document Header, June 2004. Technical Speci-
             fication Version 1.3.

[Sch88]   Willie Schatz. EDI: Putting the muscle in commerce and
             industry. Technical Report Vol. 34, Datamation, March
             1988.

[SOA03]   World Wide Web Consortium (W3C). *Simple Object*
             Access Protocol (SOAP), June 2003. Version 1.2.

[Ste94]   Steve Cook and John Daniels. *Designing Object Systems: Object-oriented modeling with Syntropy*. Prentice-Hall, 1994.

[Ste01]   Perdita Stevens. Small-Scale XMI Programming: A Revolution in UML Tool Use? Technical report, University of Edinburgh, August 2001.

[UDD02]   OASIS. *UDDI Version 2.04 API Specification*, July 2002. UDDI Committee Specification.

[UG03]   UN/CEFACT TMG. *UN/CEFACT Modeling Methodology (UMM) User Guide*, 2003. CEFACT/TMG/N093 - V20030922.

[UMa04]   Object Management Group (OMG). *Unified Modeling Language Specification, Version 1.4.2*, 2004.

[UMb04]   Object Management Group (OMG). *Unified Modeling Language Specification, Version 2.0*, 2004.

[UNC05]   United Nations. *Mandate, Terms of Reference and Procedures for UN/CEFACT*, April 2005. TRADE/R.650/Rev.4.

[Ver]   Verband der Automobilindustrie. *VDA - Verband der Automobilindustrie*.

[WS04]   Andreas Winter and Carlo Simon. Exchanging Business Process Models with GXL. In *XML4BPM 2004, Proceedings of the 1st GI Workshop XML4BPM - XML Interchange Formats for Business Process Management, Marburg (Germany)*, pages 103–122, March 2004.

[WSD01]   World Wide Web Consortium (W3C). *Web Services Description Language (WSDL)*, March 2001. Version 1.1.

[WSG04]   W3C - World Wide Web Consortium. *Web Services Glossary*, February 2004.

[WSL01]   IBM Software Group. *Web Services Language*, May 2001. Version 1.0.

[xCB03]   xCBL. *xCBL Version 4.0 Documentation*, 2003.

[XLA01]   Microsoft Corporation. *XLANG - Web Services for Business Process Design*, 2001.

[XML04]   World Wide Web Consortium (W3C). *Extensible Markup Language (XML)*, 2004. Version 1.0

[XSD04]   World Wide Web Consortium (W3C). *XML Schema*,
               2004. Version 1.0.

www.ingramcontent.com/pod-product-compliance
Lightning Source LLC
LaVergne TN
LVHW022304060326
832902LV00020B/3267